Copyright Year: 2023

Copyright Notice: by Mark Splawn. All rights reserved.

The above information forms this copyright notice: © 2023 by Mark Splawn. All rights reserved.

ISBN:978-1-304-02096-3

The PLC Programming Guide for Beginners

Book Introduction: The PLC Programming Guide for Beginners 4

Chapter 1: Introduction to PLC Programming 11

Chapter 2: Understanding the Basics of PLC 22

Chapter 3: Different Types of PLCs: An Overview 29

Chapter 4: Introduction to PLC Programming Languages 36

Chapter 5: Understanding Basic Ladder Logic 43

Chapter 6: Introduction to Structured Text Programming 49

Chapter 7: Implementing Function Block Diagram (FBD) in PLC Programming 57

Chapter 8: Sequential Function Chart: Design and Implementation 64

   Chapter 9: Fundamentals of PLC Hardware Components 73

Chapter 10: PLC Networking and Communication. 82

Chapter 11: Safety Considerations in PLC Programming 92

Chapter 12: Troubleshooting PLC Systems 102

Chapter 13: PLC Simulation and Testing 118

Chapter 14: Real-World PLC Programming Examples 130

Chapter 15: Future of PLC Programming: Trends and Predictions 136

Chapter 16: Navigating the Path to a PLC Programming Career 145

Book Introduction: The PLC Programming Guide for Beginners

Welcome to a world of automation, innovation, and limitless possibilities! In "The PLC Programming Guide for Beginners," we invite you to embark on an exciting journey into the realm of Programmable Logic Controllers (PLCs). Whether you are a curious novice, an aspiring engineer, or a seasoned technician seeking to expand your skills, this book is your gateway to understanding the fundamental concepts and practical applications of PLC programming.

Imagine a world where machines and systems work in perfect harmony, bringing efficiency, precision, and control to every aspect of our lives. PLCs lie at the heart of this technological revolution, empowering us to automate processes, monitor industrial systems, and unleash the full potential of modern automation. It is a world where mundane tasks are delegated to machines, allowing us to focus on creativity, problem- solving, and advancing the boundaries of human achievement.

As you turn the pages of this guide, you will be introduced to the captivating world of PLC programming. With each chapter, we will unravel the intricate layers of this fascinating field, guiding you from the very basics to advanced techniques. We will delve into the inner workings of PLCs, exploring the various programming languages, dissecting ladder logic, and unraveling the complexities of structured text and function block diagrams.

But this book is not just about theoretical knowledge. It is about bridging the gap between theory and practice, empowering you to apply your newfound understanding to real-world scenarios. Through vivid examples, case studies, and hands-on exercises, we will take you on a captivating journey where you can witness the transformation of concepts into tangible solutions.

As you immerse yourself in the chapters, be prepared to be challenged, to explore, and to stretch the boundaries of your imagination. PLC programming is an ever- evolving field, with new technologies, trends, and possibilities emerging at a rapid pace.

We will not only equip you with the fundamentals but also provide glimpses into the future of PLC programming, allowing you to glimpse the breathtaking possibilities that lie ahead.

Along this journey, you will discover the power of collaboration, problem-solving, and continuous learning. You will become part of a vibrant community of like-minded individuals, where knowledge is shared, ideas are nurtured, and innovation flourishes.

Embrace the excitement, the trials, and the triumphs that await you as you dive into the realm of PLC programming.

Whether you dream of designing smart factories, optimizing energy systems, or revolutionizing the way we interact with technology, "The PLC Programming Guide for Beginners" will serve as your compass, guiding you towards a future filled with endless opportunities.

Are you ready to embark on this extraordinary adventure? Let us embark together on a journey that will transform your understanding, ignite your passion, and shape your future. Welcome to the world of PLC programming!

Let the journey begin!

Chapter 1, "Introduction to PLC Programming," provides a gentle introduction to PLCs, their history, evolution, and applications in various industries.

Chapter 2, "Understanding the Basics of PLC," dives deeper into the fundamental concepts that govern PLC operations, including a detailed look at its architecture, its components, and the role it plays in industrial automation.

Chapters 3 and 4 introduce you to the different types of PLCs available in the market and explore the various PLC programming languages, setting the stage for more detailed discussions in later chapters.

Chapters 5 to 8 are dedicated to detailed discussions on PLC programming languages like Ladder Logic, Structured Text

Programming, Function Block Diagram (FBD), and Sequential Function Chart. These chapters will provide you with insights into the structure and implementation of these languages.

Chapter 9, "Fundamentals of PLC Hardware Components," delves into the physical aspects of PLCs, demystifying the hardware components that make up a PLC system.

The next set of chapters deal with advanced topics such as PLC networking and communication, safety considerations, and troubleshooting techniques.

Towards the end of the book, we will walk you through some real-world PLC programming examples (Chapter 14), and we will take a look into the future, discussing trends and predictions related to PLC programming.

This guide is a result of a rigorous effort to simplify the complex world of PLC programming and make it accessible to beginners. Whether you're an aspiring automation engineer, a professional looking to switch careers, or a curious reader wanting to explore the world of industrial automation, this book is for you.

In the spirit of PLC, let's keep learning programmable!

Chapter 1: Introduction to PLC Programming

In the realm of industrial automation, Programmable Logic Controllers, or PLCs as they are more commonly known, are the cornerstone. These robust computers are designed to control machinery and processes, carrying out specific tasks with efficiency and accuracy. But to understand their function, we must first explore the world of PLC programming.

The dawn of PLCs came with the need for a reliable and flexible solution to replace relay-based control systems in the manufacturing industry. The year was 1968, and the automotive industry was in dire need of a solution that would simplify the process of manufacturing automation. This is when Richard Morley, often hailed as the father of PLC, developed the first Modicon (Modular Digital Controller). This device, capable of being programmed to perform control tasks, marked the birth of the first

Richard Morley, Father of PLC's

PLCs soon found their place in industries far and wide, from automobile manufacturing to food processing plants, chemical factories to power plants. The reasons for their wide adoption were the many advantages they offered. They were rugged and could withstand harsh industrial environments.

They were flexible and could be reprogrammed to perform different tasks. But most importantly, they reduced the time and complexity associated with wiring relay- based control systems.

Here's an expanded timeline showcasing the key milestones in the history of Programmable Logic Controllers (PLCs) with additional details:

1968: The Birth of the PLC

- Richard Morley and his team at Bedford Associates develop the first PLC, called the "Modicon 084." It revolutionizes the automation industry by replacing complex and costly hardwired relay systems with a programmable electronic device.
- The Modicon 084 offers flexibility and ease of reprogramming, allowing manufacturers to quickly modify control logic without rewiring.

1971: PLCs in Automotive Manufacturing

- General Motors (GM) becomes one of the first adopters of PLC technology in their automotive assembly lines. By utilizing PLCs, GM significantly improves productivity, reduces downtime for reconfiguration, and

enhances the overall efficiency of their manufacturing processes.
- The successful implementation of PLCs in the automotive industry sparks widespread interest and paves the way for PLCs' adoption in other sectors.

1973: Introduction of the Allen-Bradley PLC

- Allen-Bradley, now part of Rockwell Automation, introduces their first PLC to the market. Their programmable controller offerings quickly gain popularity due to their advanced features, reliability, and user-friendly programming interfaces.
- Allen-Bradley PLCs become widely recognized for their modular design, expandability, and compatibility with various industrial applications.

1986: Standardization of PLC Programming

- The International Electrotechnical Commission (IEC) releases the IEC 61131-3 standard, establishing a unified programming standard for

PLCs. This standardization enables programmers to use common programming languages such as ladder logic, structured text, function block diagrams, and more, promoting interoperability among different PLC brands and simplifying programming efforts.
- IEC 61131-3 becomes the de facto standard for PLC programming worldwide, ensuring consistency and facilitating collaboration among programmers.

1990s: Shift towards Distributed Control Systems

- PLCs begin integrating Distributed Control System (DCS) capabilities, enabling advanced process control and distributed intelligence in industrial systems. This integration allows PLCs to operate in conjunction with other devices, such as supervisory computers and operator interfaces, to provide a holistic control solution.

- The use of DCS-equipped PLCs leads to improved operational efficiency, enhanced system scalability, and better coordination between different control units.

1996: Introduction of Ethernet Communication

- Ethernet communication protocols become common in PLCs, replacing traditional serial communication methods. The adoption of Ethernet allows for faster and more reliable data exchange between PLCs, human- machine interfaces (HMIs), supervisory control and data acquisition (SCADA) systems, and other networked devices.
- Ethernet-based communication enables seamless integration of PLCs into industrial networks, facilitating real-time monitoring, control, and data acquisition.

2000s: PLCs with Integrated HMI/SCADA

- PLC manufacturers start incorporating Human-Machine Interface (HMI) and Supervisory Control and Data Acquisition (SCADA) functionalities into their PLC systems. These integrated solutions provide a unified platform for control, visualization, and data management, reducing the need for separate HMI/SCADA hardware and software.
- PLCs with integrated HMI/SCADA capabilities offer enhanced operator interaction, advanced data visualization, and seamless integration with PLC programming environments.

2010s: Advancements in Industrial IoT and Cybersecurity

- PLCs become integral components of Industrial Internet of Things (IIoT) systems, enabling remote monitoring, predictive maintenance, and data analytics. PLCs gather and transmit valuable operational data to cloud- based platforms, facilitating real-time

insights and enabling data-driven decision-making.
- With the rise of cyber threats in industrial environments, there is a greater emphasis on PLC cybersecurity. Manufacturers implement robust security measures, such as authentication protocols, encryption algorithms, and network segmentation, to safeguard PLC systems from potential cyberattacks.

Present and Future: Industry 4.0 and Beyond

- PLCs continue to evolve in the era of Industry 4.0, with advancements in technologies such as cloud computing, artificial intelligence, and machine learning. These innovations enable smarter manufacturing processes, autonomous decision-making, and adaptive control systems.
- The role of PLCs expands beyond traditional industrial automation, finding applications in sectors such as

energy, transportation, healthcare, and more. PLCs play a crucial role in driving efficiency, sustainability, and connectivity in diverse industries, shaping the future of automation.

This expanded timeline highlights the significant milestones in the history of PLCs, showcasing their evolution from basic relay replacements to powerful and interconnected devices that have transformed industrial automation and paved the way for a new era of smart manufacturing.

At the heart of these advanced controllers is PLC programming, the process of creating instructions for the PLC to perform specific tasks. In essence, PLC programming involves writing code in a specific language understood by the PLC, which in turn performs actions based on this code.

The primary languages used for PLC programming are Ladder Logic, Structured Text, Instruction List, Function Block Diagram, and Sequential Function Chart. Each of these languages has its strengths and applications, and we will look at each in the forthcoming chapters.

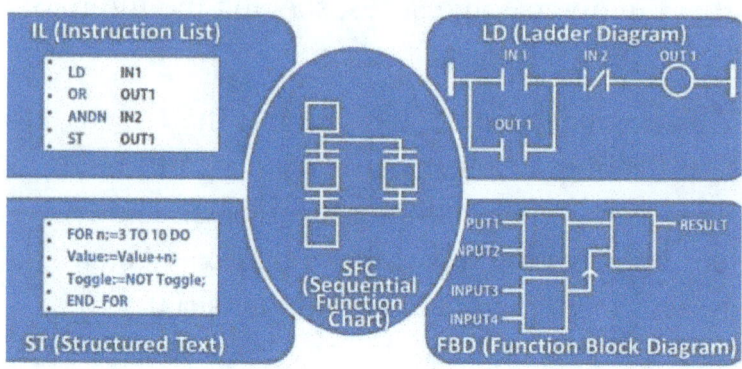

Programming a PLC begins with defining the problem, which entails understanding the task the PLC needs to perform. This is followed by planning the program, where the programmer breaks down the problem into manageable sections. Next, the programmer writes the program using a suitable PLC programming language. The program is then tested and debugged, and once it works as intended, it is loaded into the PLC.

An effective PLC program is one that not only accomplishes the task it's meant to perform but does so efficiently and safely. It should be well-structured and easily understandable to others who might need to maintain or modify it in the future.

As we delve further into this book, you will gain a deeper understanding of these concepts. The world of PLC programming is vast and fascinating, and this chapter is your first step into this exciting field.

This was a brief introduction to the world of PLC programming. The subsequent chapters will delve into more detail, unraveling the fascinating world of PLCs, and equipping you with the knowledge to understand and work with them.

Chapter 2: Understanding the Basics of PLC

PLC, or Programmable Logic Controller, is a specialized computer used in industries to control complex systems. Unlike personal computers, PLCs are designed to endure harsh industrial conditions, like heat, dust, and noise. They are also made to handle input/output arrangements in real-time, which is a requisite in control system applications.

At the core of understanding PLCs is to comprehend the basic structure and function of a PLC system. A PLC is made up of three fundamental components: a Central Processing Unit (CPU), Input/Output (I/O) modules, and a programming device.

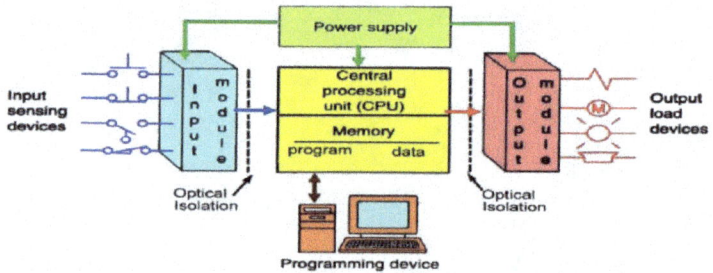

The CPU, akin to the brain in a human body, controls all PLC activities. It executes the control program that's been loaded into its memory, carries out the instructions within the program, and manages data transfers between the I/O devices and the memory. In addition, the CPU also supervises system health and status by performing self-checks and overseeing the functioning of the I/O modules.

The I/O modules are the gateways for information flow to and from the PLC system. Input modules collect information from sensing devices and send it to the CPU, and output modules receive information from the CPU to control actuators. These modules can handle numerous types and amounts of inputs and outputs, and their compatibility with the PLC CPU and the connected devices is vital for system operation.

The programming device is used to write the control program and load it into the PLC CPU. It can be a personal computer with suitable programming software, or a handheld device specifically designed for PLC programming. The control program written using the programming device defines how the PLC will perform its control tasks.

A PLC's operation can be distilled into a four-step cycle: Input Scan, Program Scan, Output Scan, and Housekeeping. During the Input Scan, the PLC checks the states of all connected input devices. In the Program Scan, it executes the user-created control program. The Output Scan involves the PLC updating the status of output devices based on the program results. Lastly, during Housekeeping, the PLC performs internal diagnostics and communication tasks.

Learning to program a PLC involves understanding these fundamental concepts and how they interact in a working PLC system. This knowledge forms the basis on which the complexities of PLC programming languages can be explored, which we'll do in the following chapters.

A PLC's **CPU** is built around a microprocessor, which performs the bulk of the processing tasks. The CPU is equipped with a memory system that's divided into two categories - user memory and system memory. The user memory stores the PLC program and the data associated with the program, while the system memory is used by the PLC's operating system for internal operations and communication.

The **Input/Output (I/O) modules** are the link between the PLC and the physical world. Input modules read signals from sensing devices like switches, temperature sensors, or light sensors, and convert them into a digital form that the CPU can interpret. They are essentially the eyes and ears of the PLC. On the other hand, output modules convert the CPU's digital signals into a form that can control actuators, such as motors, lights, or valves, allowing the PLC to interact with its environment.

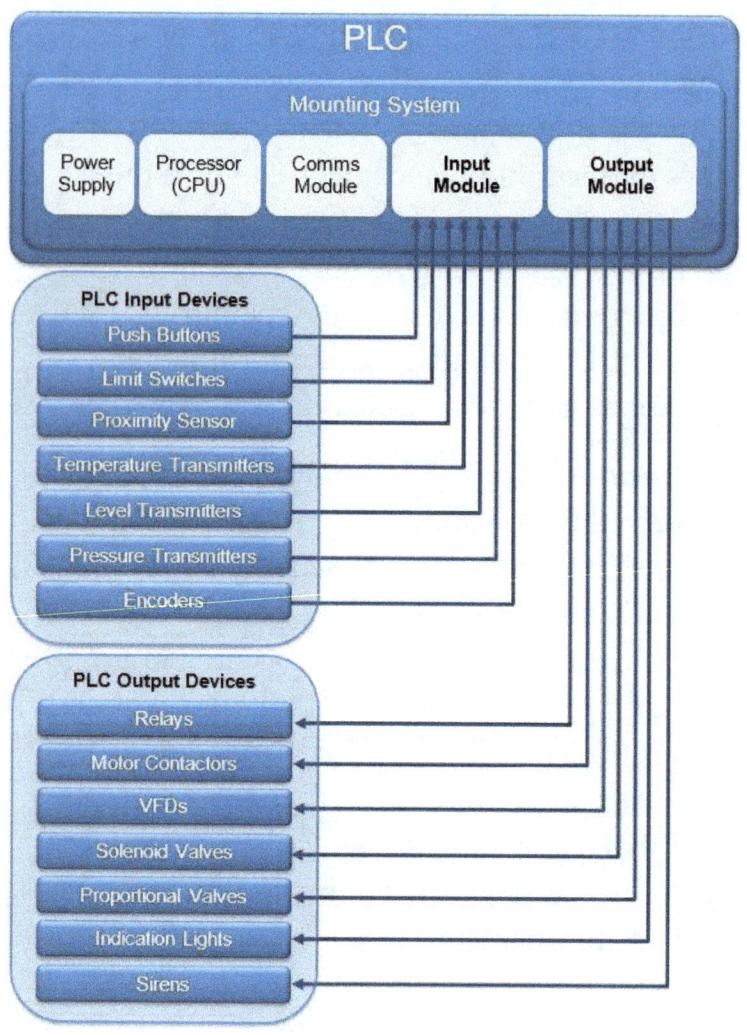

The choice of I/O modules depends on the specific needs of the control system, such as the types of devices it will interact with and the nature of those interactions.

Understanding the specific needs of your

control system is crucial for selecting the appropriate I/O modules.

The **programming device** is used to write, edit, and debug the control program. This device can be a computer equipped with specialized software, a dedicated handheld programmer, or even a smartphone or tablet with appropriate apps. The choice of programming device often depends on the programmer's preferences, the complexity of the control program, and the specific PLC system being programmed.

PLC operation involves a repeated cycle of four steps. During the **Input Scan**, the PLC reads the current states of all input devices and stores this information in its memory. During the **Program Scan**, the PLC executes the control program one instruction at a time. Each instruction's result may depend on the input states, and the results are stored in memory for use by later instructions.

During the **Output Scan**, the PLC updates the states of all output devices based on the results of the control program. Finally, during **Housekeeping**, the PLC performs self-checks to ensure it's functioning correctly and manages any communication tasks.

The underlying logic of PLC operation is a reflection of the real-world industrial processes it controls. In a typical process, the states of various devices are first monitored (Input Scan), then decisions are made based on these states (Program Scan), and finally, actions are carried out based on these decisions (Output Scan).

The upcoming chapters will delve deeper into PLC programming languages and how you can use them to define the PLC's control tasks. Stay tuned as we embark on this journey to explore the fascinating world of PLC programming!

Chapter 3: Different Types of PLCs: An Overview

As we venture deeper into the realm of Programmable Logic Controllers, it's vital to appreciate that not all PLCs are created equal. Various types of PLCs are available on the market, each tailored to specific needs and applications. Understanding these differences can help you choose the right PLC for your project or even help you better understand the PLC you're already working with.

PLC
Programmable Logic Controller

The fundamental categorization of PLCs is based on their sizes and capabilities, leading to three main types: Nano PLCs, Micro or Compact PLCs, and Modular or Rack-Mounted PLCs.

Nano PLCs are the smallest form of PLCs available in the market, and as their name suggests, they are compact in size but are mighty in function for their size. Designed with simplicity in mind, Nano PLCs have less than 100 I/O points. They are typically not expandable due to their limited CPU capacity. Despite their limitations, Nano PLCs have found their niche in small-scale applications such as home automation systems, vending machines, and small conveyors. They are the perfect choice when the control task is simple, and the budget is tight. Besides, their small footprint makes them ideal for situations where space is a constraint.

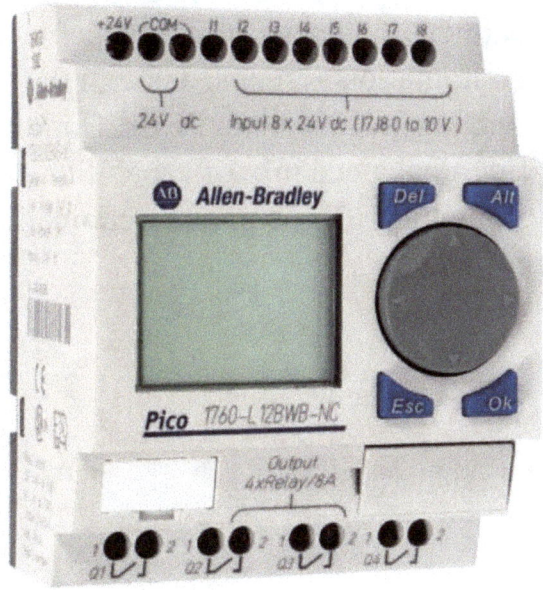

Taking a step further, we have **Micro or Compact PLCs**. These PLCs are more complex than Nano PLCs and can handle a few hundred I/O points. Their capabilities often extend to features such as data logging, communication protocols like Ethernet or Profibus, and limited expandability. Compact PLCs are commonly used in applications that require a moderate level of control complexity and more I/O points than a Nano PLC can handle. Some examples include water treatment plants, packaging machines, and building automation systems. Though they are costlier and larger than Nano PLCs, Compact PLCs offer a balanced mix of capability, size, and price, making them a popular choice for many applications.

Finally, at the top of the PLC hierarchy, we have **Modular or Rack-Mounted PLCs**. These PLCs are the most flexible and capable, designed to manage thousands of I/O points. They can be expanded by adding more modules, allowing them to adapt to a wide range of complex applications. With features such as advanced communication capabilities, PID control, motion control, and data manipulation, Modular PLCs cater to high-demand applications. These include oil refineries, power plants, and large manufacturing lines. While these PLCs come with a higher price tag and larger size, the customization and expandability they offer make them well worth it for large-scale applications.

Understanding the differences between these PLC types is crucial in selecting the right PLC for your needs. Consider factors like the number of I/O points, required features, available space, and your budget when making your decision. Remember, the most expensive or complex PLC isn't necessarily the best choice. The best PLC for your application is one that meets your specific needs without being overkill.

This chapter has provided a broad overview of the different types of PLCs, offering a glimpse into their applications and strengths. As we delve deeper into the world of PLC programming, we'll examine how these differences can impact the way we approach programming tasks.

Chapter 4: Introduction to PLC Programming Languages

With an understanding of the different types of PLCs and their functionalities, we can now dive into the essence of PLCs – their programming languages. PLCs are programmed using a variety of languages, each with its unique strengths and suitable applications. The International Electrotechnical Commission (IEC) has standardized five types of PLC programming languages under the IEC 61131-3 standard. They include: Ladder Diagram (LD), Instruction List (IL), Structured Text (ST), Function Block Diagram (FBD), and Sequential Function Chart (SFC).

Ladder Diagram (LD) is a favorite among many PLC programmers, especially those with a background in electrical or control engineering. This is because LD's graphical representations closely resemble electrical relay control circuit diagrams. The symbols represent contacts and coils, where a contact can be thought of as an input or condition, and a coil as an output or action. LD's power lies in its simplicity and visual nature, allowing programmers to easily understand and design control logic.

For example, if we want to control a motor based on the states of two switches, this could be represented in an LD program as two contact symbols in series (representing the two switches) connected to a coil symbol (representing the motor). This would mean that the motor turns on only when both switches are on.

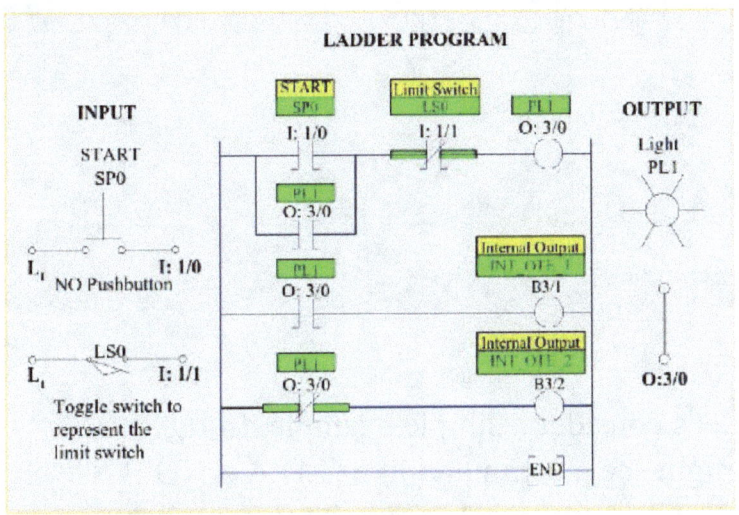

Instruction List (IL), also known as Statement List (STL), is a textual language that might seem cryptic to those unfamiliar with assembly or machine languages.

Despite being less human-readable, IL is a compact and efficient language that can be highly effective once you become comfortable with it.

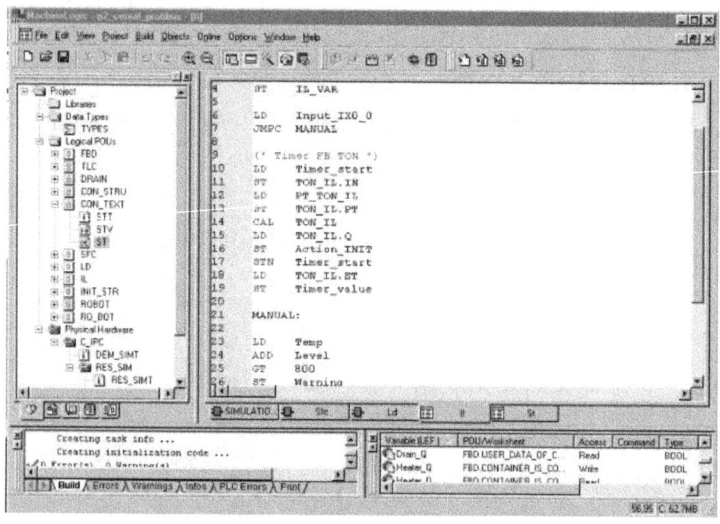

Let's consider a simple example. In IL, you might see a command like "LD A AND B ST Q," where A and B are inputs (like switches), and Q is an output (like a motor). This would mean, load the value of A, and if A is true, also check the value of B, and if both are true, set Q to be true.

Structured Text (ST) is a high-level textual language that looks a lot like Pascal or Basic. Programmers comfortable with traditional computer programming languages might find ST more familiar and flexible. It's great for more complex mathematical operations or algorithmic tasks that might be awkward to represent graphically.

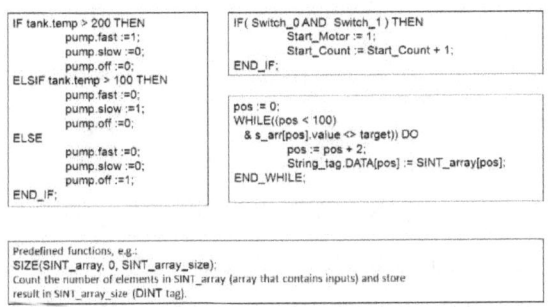

An example in ST could be "IF A AND B THEN Q:= TRUE; ELSE Q:= FALSE; END_IF,"
which essentially performs the same function as the previous examples in LD and IL but

written in a more traditional programming format.

Function Block Diagram (FBD) represents control systems as interconnected blocks, each representing a function or operation. This graphical language is a favorite for complex control systems, where it's helpful to visualize the relationships between different elements.

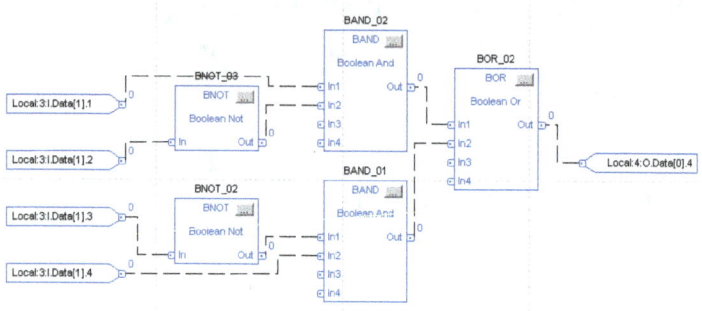

For instance, if you have two temperature sensors providing inputs to a heater (with a complex control algorithm), you might have two input blocks representing the sensors, an algorithm block that takes these inputs and calculates the required heater state, and an output block representing the heater.

Sequential Function Chart (SFC) is another graphical language that is particularly useful for processes that involve a sequence of operations. Each step in an SFC program represents a specific state of the system, and the transitions between the steps represent the conditions that need to be met to move from one state to another.

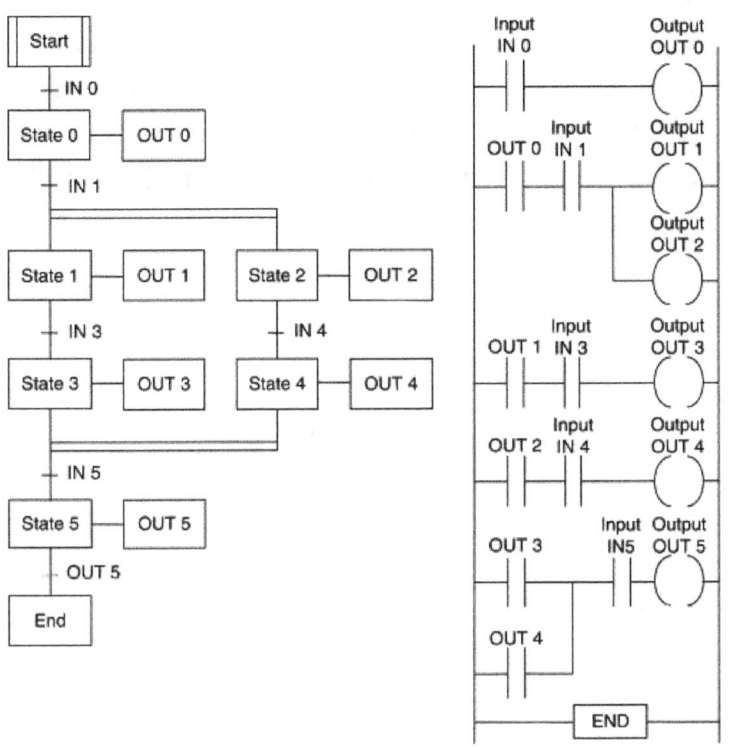

An example in SFC might be a conveyor system where items need to be washed, dried, and then packed. Each of these operations would be a step, and the conditions to transition from one operation to the next could be based on timers, sensors, or other inputs.

Remember, each of these languages has its strengths and is better suited to some tasks than others. The best PLC programmers often have a working knowledge of all of them and can choose the best one for a particular application.

Chapter 5: Understanding Basic Ladder Logic

Now that we have a broad overview of the five programming languages, we will delve deeper into the most prevalent and user- friendly one - the Ladder Diagram (LD). This chapter is dedicated to understanding the basic elements and operations of ladder logic.

The structure of a ladder logic program resembles a ladder, with two vertical rails and several horizontal rungs. The left rail is connected to a power source, and the right rail is connected to the ground. The rungs in between contain the control logic.

There are three basic components in a ladder logic diagram: contacts, coils, and lines.

Contacts represent the inputs and can be normally open (NO) or normally closed (NC). A normally open contact will close (or turn ON) when its corresponding input is energized. Conversely, a normally closed contact will open (or turn OFF) when its corresponding input is energized.

Coils represent the outputs. When a coil is energized, its corresponding output is turned ON, and when it is de-energized, the output is turned OFF.

Lines are the horizontal lines that connect the contacts and coils. When a line or path can be traced from the left rail (power source) through closed contacts to a coil and then to the right rail (ground), the coil is energized.

Basic logic operations can be performed in ladder logic using these elements. For instance, an AND operation is represented by connecting contacts in series, while an OR operation is represented by connecting contacts in parallel.

Let's consider a simple example. Suppose we have two switches, A and B, controlling a light bulb, L. If we want the light to turn on when both switches are ON (AND operation), we place contacts A and B in series on a rung, with the coil representing L. If we want the light to turn on when either switch is ON (OR operation), we place contacts A and B in parallel.

Remember, this is a simplistic representation of ladder logic operations. Actual PLC systems may involve complex combinations of these operations and various types of special function blocks.

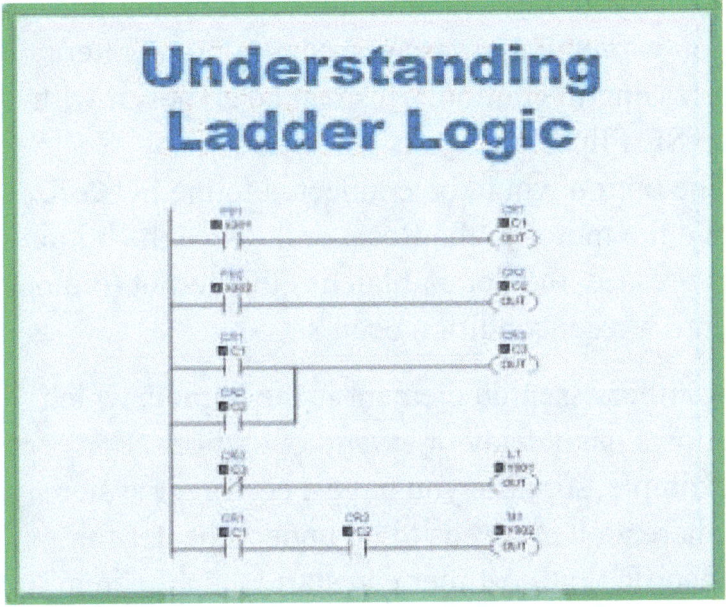

As previously mentioned, Ladder Logic employs a graphical language reminiscent of electrical relay logic diagrams, making it intuitive for people with a background in electrical engineering. However, there's more complexity to Ladder Logic than initially meets the eye. Let's discuss some additional elements.

One such element is the **Set/Reset (SR) Flip Flop**. This is a control circuit used to maintain a state, even when the actuating condition is no longer present. Suppose you have a system where you press a button to start a motor, and you want the motor to continue running after releasing the button. This can be achieved with an SR Flip Flop.

The button would be connected to the Set coil, and the motor to the Reset coil. Once the button is pressed, the Set coil latches the output (motor) until a reset condition occurs.

Another essential element is the **Timer**, which allows us to delay or extend operations. For example, suppose you have a conveyor system where an item needs to be under a heater for a specific time. A timer can start when the item reaches the heater (start condition) and stop the conveyor when the set time has elapsed.

Counters are useful when we want to perform an operation after a certain number of occurrences. For example, a counter can be used in a packaging system to trigger a sealing operation after a set number of items have been packed.

Ladder Logic also allows for **Arithmetic Operations**. Suppose a tank system uses two sensors (Sensor A and Sensor B) to measure the level of a liquid. The control system can calculate the average of the two sensor readings and use it to control a pump.

In a Ladder Diagram, these elements can be represented by special function blocks. Their exact graphical representation and use can differ between different PLCs, but the underlying concept remains the same.

For example, a simple Ladder Logic program might look like this:

1. Start condition (Start Button) – Normally Open (NO) contact
2. Stop condition (Stop Button) – Normally Closed (NC) contact
3. Output (Motor) – Coil

This represents a basic motor control circuit where the motor starts when the Start Button is pressed and stops when the Stop Button is pressed.

Remember, real-world PLC applications can be much more complex, involving nested operations, multiple function blocks, and numerous I/O points. Mastering Ladder Logic requires understanding these basic elements and then practicing combining them to solve control problems.

In the next chapters, we will dive deeper into these elements and more complex Ladder Logic concepts.

Chapter 6: Introduction to Structured Text Programming

Structured Text (ST) is one of the five PLC programming languages defined by the IEC 61131-3 standard. Unlike Ladder Logic, which is graphical, Structured Text is a high- level, text-based language similar to Pascal or Basic. The text-based nature makes ST suitable for complex algorithms or mathematical operations that may be awkward to represent graphically.

ST is block-structured, which means that the code is divided into individual blocks, each responsible for a specific task. These blocks can be Functions, Function Blocks, or Programs. Each block contains variables (inputs, outputs, in-outs, and locals) and instructions.

PLC Controls with Structured Text (ST)

```
//Check level low - level cannot be negative
IF Level < 0 THEN
    Level:= 0;
END_IF;

//Check level high - tank cannot be overfilled
IF Level > (TankRadius + TankHeight) THEN
    Level:= TankRadius + TankHeight;
END_IF

//Half circle ball
IF Level <= TankRadius THEN
    Lr:= SQRT(Level * (TankDiameter - Level));
    Vol:= (PI/6)*level*(3*Lr*Lr+ Level*Level);
ELSE
    //Half circle ball filled
    Vol:= 2/3*PI * TankRadius * TankRadius * TankRadius;
END_IF;

//Something in the cylinder
IF Level > TankRadius THEN
    Vol:= Vol + (Level - TankRadius) * PI * TankRadius * TankR
END_IF;
```

The instructions are written in lines and executed sequentially, although control structures like IF, CASE, FOR, WHILE, and REPEAT can alter the flow of execution. Comments can be added to the code using (* *) or //, which help in understanding and maintaining the code.

ST supports a wide range of operators and functions, including:

1. Arithmetic Operators: +, -, *, /, MOD (modulo), and ** (exponentiation).

2. Comparison Operators: = (equal), <> (not equal), < (less than), <= (less than or equal), > (greater than), and >= (greater than or equal).
3. Logical Operators: AND, OR, XOR, and NOT.
4. Bitwise Operators: AND, OR, XOR, and NOT.

To illustrate the use of ST, let's consider a simple example of a pump control system. Suppose we have a tank with a level sensor and a pump. We want the pump to turn on when the level drops below 20% and turn off when it reaches 80%.

This could be implemented in Structured Text as follows:

IF Level <= 20 THEN

  Pump := TRUE;

ELSIF Level >= 80 THEN

  Pump := FALSE;

END_IF

(IF Level <= 20 THEN Pump := TRUE; ELSIF Level >= 80 THEN Pump := FALSE; END_IF)

Here, Level represents the input from the level sensor, and Pump represents the output to the pump. ":=" is used for assignment, and ";" denotes the end of a statement.

The IF statement checks if the Level is less than or equal to 20. If it is, it sets the Pump to TRUE, turning it on. If the Level is not less than or equal to 20, it checks the ELSIF condition. If the Level is greater than or equal to 80, it sets the Pump to FALSE, turning it off.

This is a very basic example, but it illustrates the fundamental structure and operation of Structured Text. As we delve deeper into ST in the following chapters, you'll learn to handle more complex situations and achieve more sophisticated control.

Structured Text (ST) programming is extensively used when we need to solve complex problems or perform operations that involve heavy computation, which are difficult to achieve with graphical languages like Ladder Logic.

The high-level, text-based nature of ST allows programmers to implement complex algorithms, mathematical calculations, data manipulations, and handle a wide array of control tasks. These capabilities make ST particularly useful for the control of processes where intricate logic and calculations are involved, such as in advanced manufacturing or chemical processes.

To explain further, let's take a look at some of the fundamental constructs of ST and their applications:

1. Variables: In ST, you can declare and use different types of variables, each having specific characteristics. You have access to standard data types, including but not limited to BOOL (Boolean), INT (Integer), REAL (floating point numbers), and STRING (text strings). Custom data types can also be created according to the application's requirements.

For example, consider a temperature control system where you need to read the temperature value (a REAL type variable) from a sensor, compare it with a set point (another REAL type variable), and control a heater (a BOOL type variable).

2. Control Structures: ST offers various control structures to guide the flow of the program, including conditional statements (IF, CASE) and looping structures (FOR, WHILE, REPEAT).

For example, in a packaging system where different actions are needed for different product types, a CASE structure could be used. In a system where an action needs to be repeated a certain number of times, a FOR loop could be implemented.

3. Functions and Function Blocks: ST supports the use of standard functions (like SIN, COS, SQRT, LOG, etc.) and function blocks (like timers and counters). It also allows the creation of custom functions and function blocks, which can then be reused throughout the program.

For instance, in a system that requires computation of complex mathematical equations, standard functions can be used. For a system requiring timing operations, timer function blocks can be deployed.

4. Comments: In ST, comments can be added using (* *) or // to make the program easier to understand and maintain.

Taking the pump control system example further, suppose we also have a heater in the tank to maintain the water temperature. We want the heater to turn on when the temperature drops below 15 degrees and turn off when it reaches 25 degrees. This could be implemented in Structured Text as follows:

IF Level <= 20 THEN

  Pump := TRUE;

ELSIF Level >= 80 THEN

```
    Pump := FALSE;
END_IF


IF Temp <= 15 THEN
   Heater := TRUE;
ELSIF Temp >= 25 THEN
   Heater := FALSE;
END_IF
(IF Level <= 20 THEN Pump := TRUE; ELSIF
Level >= 80 THEN Pump := FALSE; END_IF
IF Temp <= 15 THEN Heater := TRUE; ELSIF
Temp >= 25 THEN Heater := FALSE; END_IF)
```

Here, Temp represents the input from a temperature sensor, and Heater represents the output to the heater.

In the following chapters, we will dive deeper into the specifics of Structured Text programming and also learn about other PLC programming languages.

Chapter 7: Implementing Function Block Diagram (FBD) in PLC Programming

The Function Block Diagram (FBD) is another key language defined in the IEC 61131-3 standard for PLC programming. This language, like Ladder Logic, is graphical, but its representation and usage are quite different.

In FBD, the program is built by linking function blocks together in a block diagram format. This format is often used when the logic involves complex control tasks, especially when the task is more data flow oriented rather than input/output oriented.

A Function Block is a pre-programmed instruction that performs a specific function in the control process, such as logical operations, mathematical operations, timers, counters, and data manipulation. These blocks have inputs and outputs and can be linked together to form a complete control program.

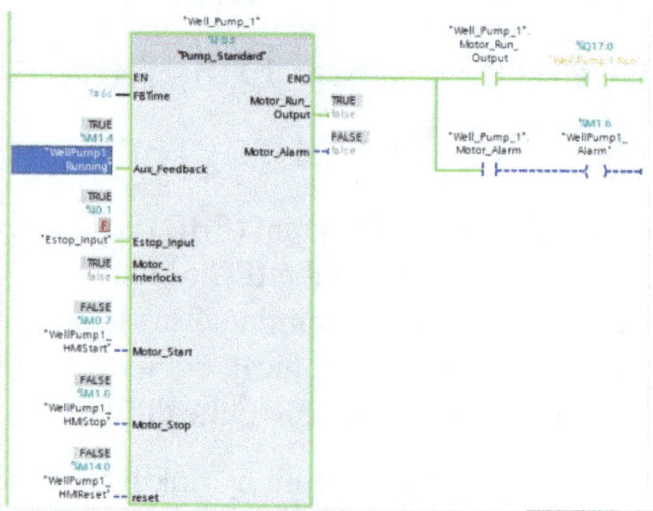

The primary building blocks of an FBD program are:

1. Standard Functions (SF): These are the fundamental logic gates (AND, OR, NOT, XOR) and arithmetic operations (ADD, SUB, MUL, DIV).

2. Standard Function Blocks (SFB): These are complex instructions like timers, counters, and PID controllers. They maintain internal state information and thus, can be used for operations that need to remember past states or events.

3. User-defined Function Blocks (UFB): These are custom function blocks defined by the user to suit specific application needs.

In an FBD program, a function block is represented as a box with inputs entering from the left and outputs exiting to the right. The boxes are connected by lines, called wires, to pass data between them.

Let's consider a simple example: suppose we have a pump and two start conditions, Start1 and Start2. The pump should start if either Start1 or Start2 is true, and there is a Stop condition that can stop the pump.

This could be implemented in FBD as follows:

- Use an OR block to represent the logic between Start1 and Start2.
- Connect Start1 and Start2 to the inputs of the OR block.
- Use an AND block to represent the logic between the result of the OR block (Start1 OR Start2) and the Stop condition.

- Connect the output of the OR block and the Stop condition to the inputs of the AND block.
- Connect the output of the AND block to the Pump.

The advantage of FBD is that it allows for easy visualization and understanding of the control logic, especially for complex control tasks. Furthermore, with the use of function blocks, repetitive tasks can be easily modularized and reused, reducing programming effort.

In the next chapter, we will dive deeper into creating custom function blocks and using advanced features of the FBD language.

FBD is often regarded as one of the most intuitive and accessible PLC languages due to its graphical nature and its reliance on pre- built blocks that can easily be connected to form a control strategy. Here, we will look at the important components of the FBD language and use examples to clarify their use.

1. Boolean Operations in FBD: The fundamental logical operations in FBD are AND, OR, and NOT. These basic blocks can be linked together to form any logic expression.

For instance, consider a system where a conveyor belt needs to start when either of two start buttons (Start1 or Start2) is pressed, but can be stopped by a stop button. We can implement this using two OR blocks (one for the start buttons and another for the stop button along with the OR block's output) and a NOT block for the stop button.

2. Timers and Counters: Like other PLC languages, FBD also has built-in timer and counter blocks. These blocks can be connected in the diagram to perform timing and counting operations.

Consider a situation where a conveyor belt needs to move for a specific time after a start button is pressed. This could be achieved using a timer block, with the start button connected to the 'IN' input of the timer, and the 'Q' output of the timer controlling the conveyor.

3. Numeric Operations: FBD supports a wide range of numeric operations, including addition, subtraction, multiplication, division, and modulus. These blocks can take one or more numeric inputs and produce a numeric output.

For instance, suppose we want to control a mixer motor's speed based on two factors: the viscosity of the material and the quantity to be mixed. We could use multiplication and addition blocks to calculate the desired speed based on these two factors.

4. Comparison and Select Blocks: These blocks allow for comparison of two values and can be used for making decisions. The Select block can be used to choose between two values based on a condition.

For example, in a temperature control system, a comparison block can be used to compare the current temperature with the setpoint. The output of this block could then be used with a Select block to choose between two different heater states (on/off).

5. User-defined Function Blocks (UFB): Apart from using the standard function blocks, FBD allows the creation of User- defined Function Blocks. These blocks can encapsulate a part of the control logic that can be used multiple times throughout the program, promoting reusability and maintainability.

For example, if a certain complex computation is performed in several places in the program, it can be put into a UFB and then simply inserted wherever needed.

These are the key elements of Function Block Diagram programming. However, the true strength of FBD comes from the ability to freely combine these elements, resulting in highly flexible and powerful control logic. As we move further, we will discuss more complex examples and advanced FBD programming techniques.

Chapter 8: Sequential Function Chart: Design and Implementation

The Sequential Function Chart (SFC) is yet another method of programming PLCs. Just like Ladder Logic, Structured Text, and Function Block Diagram, SFC is a standard programming language outlined in the IEC 61131-3 standard. As the name implies, Sequential Function Charts focus on the sequence of operations in a process control system, making it an excellent choice for programming complex systems and tasks that include various stages or steps.

SFC is a graphical language that combines elements of flowcharting with temporal control methodologies, particularly for orchestrating the execution of other types of code such as ladder logic or structured text. It's often used to structure the internal organization of a program, as it provides a high-level view of the control strategy.

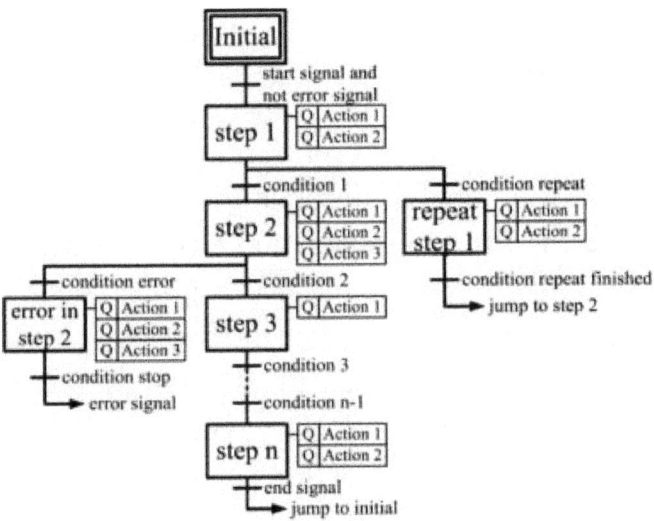

The primary components of an SFC program are steps, transitions, and actions. Let's take a closer look at each of these components:

The real beauty of SFC lies in its visual nature that provides programmers with a quick overview of the process flow. Its distinct feature is that it focuses more on the order of operations rather than the operations themselves. This makes SFC an exceptional choice for applications involving process control, batch processing, and systems that require complex sequences or stages.

Understanding Steps, Transitions, and Actions Further

1. Steps: A 'step' in SFC represents a condition or a state in the process. For example, in a manufacturing process, a step could represent a machine being idle, running, or in maintenance mode. Steps are visually represented by rectangles or squares in the SFC diagram. Each step in an SFC diagram may have one or more associated actions. These actions are only executed when the step is active.

A specific type of step, known as an 'Initial Step', is the entry point of the program.
There is only one Initial Step in an SFC program and it is activated when the PLC starts.

2. Transitions: Transitions are conditions that allow the process to progress from one step to the next. They are represented by horizontal lines between steps in the SFC diagram. Transitions can be associated with boolean expressions; when the expression is evaluated as 'True', the transition is enabled.

For example, a transition in a manufacturing process could be "Has the machine finished the current task?". If the answer is 'True' (yes, the machine has finished the current task), the transition condition is satisfied, and the program moves to the next step.

3. Actions: Actions in SFC are tasks performed during a step. They are typically represented by a rectangle with a dashed outline and can be written in other PLC languages such as Ladder Logic, Structured Text, or FBD. Actions are associated with steps and are only executed when their associated step is active.

For instance, an action during a 'Running' step might be to control the motor of a machine. If the 'Running' step is active, the action to control the motor is executed.

Sequential Function Chart Example

Consider a simplified production process in a factory, where products are first assembled, then tested, and finally packed. If we were to program this process using SFC, we would create three steps: 'Assembly', 'Testing', and 'Packing'. Each of these steps would have actions associated with them that control the respective machines for assembly, testing, and packing.

The transitions in this example could be conditions like 'Assembly completed', 'Testing completed', etc. These transitions ensure that the next step in the process isn't started until the current one is completed.

To summarize, SFC provides a high-level view of a control system in terms of steps and transitions between those steps. This view is intuitive and mirrors the way that we often conceptualize processes. In the next chapters, we will explore more complex applications of PLC programming and delve into more advanced topics.

Advantages and Limitations of SFCs

SFCs, with their visual, flowchart-like layout, are particularly beneficial in certain situations. First, they are excellent for depicting process control systems that have a strong sequence of operations or a procedural flow. If you need to develop a control program for a multi-station assembly line or a batch processing system, for example, an SFC could be an ideal choice.

Second, SFCs inherently support parallelism. That is, they can represent multiple simultaneous sequences of operations. If you're dealing with a system that requires the coordination of multiple concurrent tasks, SFCs can elegantly express such control logic.

Lastly, SFCs are great for handling exceptions and errors. By incorporating alternative paths into the chart, you can design the program to handle errors and recover from them.

Despite their advantages, SFCs also have certain limitations. For simple control tasks, using SFCs could overcomplicate the program. They might also not be the best fit for systems that don't have a clear sequential or procedural flow. And although they do support concurrent sequences, SFCs might not be ideal for systems that require complex synchronization or communication among parallel sequences.

Deep Dive into SFC Actions

As mentioned before, an SFC action is a task that's performed during a step, which can be written in other PLC languages such as Ladder Logic, Structured Text, or FBD. It's important to note that actions in SFCs are more flexible than they might first appear.

There are different types of actions that execute under different circumstances. Some of the common types of actions are:

1. **Non-reentrant actions**: These are actions that must be completed once started, meaning they can't be interrupted by another instance of the same action.
2. **Reentrant actions**: These are actions that can be interrupted by another instance of the same action.

3. **Background actions**: These actions continue to run in the background until the step they're associated with is active.

Understanding the different types of actions in SFCs and when to use them is crucial in creating effective PLC programs using this methodology.

Exploring SFCs in Real-World Scenarios

Consider a car manufacturing process. The main steps could include 'Chassis Assembly', 'Body Installation', 'Painting', and 'Final Inspection'. Each step would have multiple actions controlling different machines or robots.

Transitions would check whether each step was successfully completed, and if any errors occurred, alternative paths could handle those situations, possibly triggering alarms or rerouting the process. This makes the control program robust and resilient, thereby ensuring smooth operations.

Remember that SFC is just one of the IEC 61131-3 standard PLC programming languages. The choice of programming language depends largely on the requirements of the specific application. While ladder logic may be suitable for certain applications, SFCs may be a better choice for others.

In the upcoming chapters, we will take a closer look at the components that make up a PLC and understand how they work together to perform control tasks.

Chapter 9: Fundamentals of PLC Hardware Components

A Programmable Logic Controller (PLC) is far more than just a piece of software. It's an industrial computer control system that continuously monitors the state of input devices and makes decisions based upon a custom program to control the state of output devices. This might sound complex but bear with me - this chapter will help break down and demystify the core hardware components of a PLC, making it easier to understand its operation.

PLC Processor Unit (CPU) - A Closer Look

The PLC Processor Unit is essentially the "brain" of the entire PLC system. This is where your written PLC program is processed and executed. The CPU interprets the input signals based on the user's program, performs the operations and sends corresponding output signals.

Consider the CPU like a chef in a restaurant. It takes the raw ingredients (inputs), processes them according to the recipe (user program), and produces the final dish (outputs). The efficiency and speed of the CPU play a crucial role in how effectively the PLC can perform its tasks.

Expanding on Input and Output Modules

When we talk about inputs and outputs in the context of PLCs, we are referring to the physical connections that link the PLC to the machinery or process that it is controlling.

Let's imagine a PLC that controls an automated car wash. The input modules could be connected to sensors that detect the presence of a car, the type of wash selected, etc. Based on these inputs, the PLC program decides what operations to perform
- activate water jets, dispense soap, operate brushes, and so on. These operations are carried out by devices connected to the output modules.

The robustness of the input and output modules is crucial. As they interface with the real world, they must be able to withstand the operating conditions of the environment they are in.

Delving Deeper into the Power Supply

The power supply unit's main job is to convert the available power (which could be AC or DC) into a form that the PLC can use. Most PLCs operate on a DC voltage, but some PLCs also accept AC voltage.

In the automated car wash example, suppose the PLC operates at 24V DC, but the local power supply is 120V AC. The power supply unit would convert the 120V AC into 24V DC that the PLC can safely and efficiently use.

Understanding Memory in More Detail

The memory of a PLC is divided into several areas, each with a specific role:

1. **Program Memory**: This is where your PLC program is stored.

2. **Data Memory**: This stores information from your I/O modules and any intermediary data from your program.
3. **Status Memory**: This stores information about the PLC itself, such as error statuses and system information.

Communications Ports and Their Importance

The communication ports are crucial in today's interconnected world. They allow the PLC to exchange data with other devices like computers, other PLCs, and advanced industrial equipment. Using these ports, PLCs can be networked together to control more complex systems. They can also be connected to Human-Machine Interface (HMI) devices for easier monitoring and control.

This concludes a deeper exploration of the key hardware components of a PLC. Remember that these components can vary significantly depending on the specific make and model of the PLC. Always refer to the manufacturer's documentation for precise information about a specific PLC.

In the next chapter, we'll see how all these components come together when PLCs are networked and communicate with each other.

Rack or Chassis

The rack or chassis is essentially the 'body' of the PLC, housing the various components that we've discussed. It provides structural support, holds the modules together, and allows for data and power to be passed between them. Depending on the size and requirements of the PLC system, a rack can have different numbers of slots for accommodating various modules.

For instance, in our automated car wash scenario, a compact rack might be used due to the limited number of inputs (sensors) and outputs (water jets, brushes, etc.). In contrast, a large manufacturing plant might require a much larger rack to accommodate a greater number of I/O modules, communication modules, and special function modules.

Special Function Modules

Apart from the standard components, a PLC can also incorporate special function modules based on the application's requirements. These modules extend the capabilities of the PLC system and can be used for a wide variety of functions.

For example, a high-speed counter module can be used to track events that occur too quickly for the standard I/O modules to register. In a production line scenario, this could be used to accurately count the number of products passing a certain point.

Other examples of special function modules include analog-to-digital (ADC) and digital- to-analog (DAC) converters, temperature sensing modules, and communication modules for specific industrial protocols.

Human-Machine Interface (HMI)

While not a component of the PLC itself, the Human-Machine Interface (HMI) plays a crucial role in most PLC systems. The HMI allows for interaction between the human operator and the PLC system. Through the HMI, the operator can monitor the system's status, input commands, or adjust system parameters.

In our car wash example, the HMI could be a simple panel with buttons for selecting the type of wash and displays showing the status of the wash. In a more complex industrial setup, the HMI could be a sophisticated computer system running specialized software for monitoring and controlling the entire operation.

PLC Selection Considerations

Choosing the right PLC for your application involves understanding these hardware components and how they will interact with your specific requirements. Some key considerations include:

1. **Number of I/O Points**: Determine the total number of input and output points required for your application. Remember to consider potential future expansions.
2. **Type of I/O Points**: Understand if you require digital or analog I/O points, or perhaps a combination of both.
3. **Communication Requirements**: Does your PLC need to communicate with other devices or systems? If so, what type of protocol do they use?
4. **Environment**: Consider the operating environment and select a PLC that can withstand those conditions. For

instance, will the PLC be exposed to extreme temperatures, dust, or moisture?

5. **Budget**: PLCs can range widely in price depending on their capabilities. It's important to balance between the requirements and the available budget.

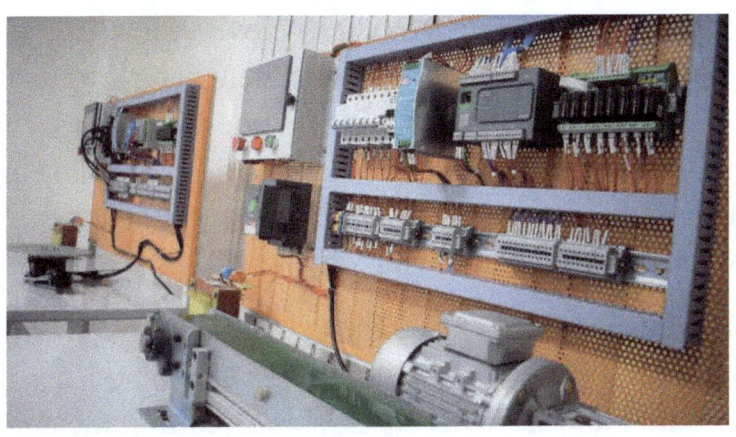

Having a thorough understanding of these hardware components and considerations will significantly aid you in programming and troubleshooting PLC systems. It will also help you make the best choices for the applications you work on.

In the next chapter, we'll explore how these PLCs can be networked together for more complex control scenarios.

Chapter 10: PLC Networking and Communication.

The Need for Networking in PLC Systems

With modern industries leaning towards automation, PLC systems are growing in complexity. It's no longer sufficient for a single PLC to control a system independently. Today, multiple PLCs often work together, networking and communicating to control larger and more complex systems.

For example, in a large factory, you might have different PLCs controlling different parts of the production line. These PLCs must communicate with each other to ensure smooth operation of the entire factory.

Different Types of PLC Networks

There are several ways PLCs can be networked together:

1. **Star Network**: In this network, each PLC is connected to a central hub. The

hub routes the communication between different PLCs.
2. **Ring Network**: Here, each PLC is connected to two others, forming a ring. Information is passed from one PLC to the next around the ring.
3. **Bus Network**: In a bus network, all PLCs are connected to a single communication line, or bus.

The choice of network depends on the specific requirements of your PLC system.

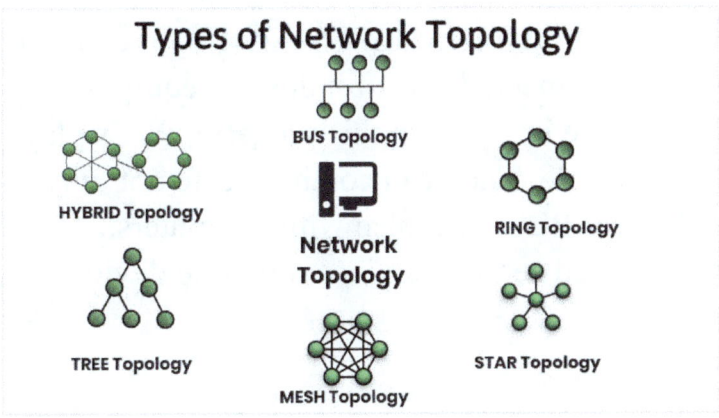

PLC Communication Protocols

PLCs use specific communication protocols to exchange information. These protocols define the rules for how data is transmitted and received. Some of the most common protocols used in PLC systems include Modbus, Profibus, and Ethernet/IP.

Each protocol has its advantages and disadvantages, and the choice of protocol can significantly impact your PLC system. For example, Modbus is simple and easy to implement but lacks the speed and features of more modern protocols like Ethernet/IP.

The Role of PLCs in Industrial Internet of Things (IIoT)

The Industrial Internet of Things (IIoT) refers to the interconnectedness of industrial equipment through the internet. In this scenario, PLCs play a crucial role. They can communicate not only with other PLCs, but also with computers, cloud-based servers, and even mobile devices.

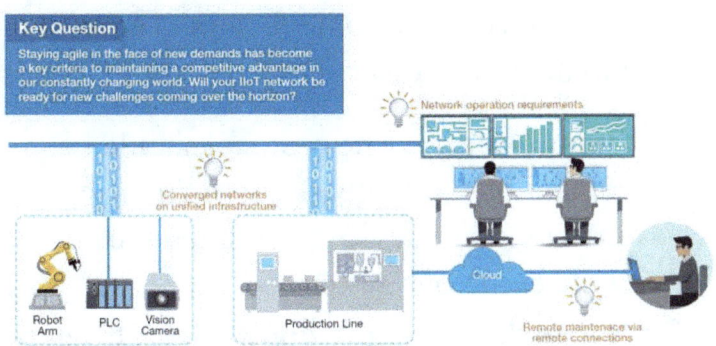

With IIoT, a factory manager could monitor and control the production line remotely from a laptop or smartphone. This brings an entirely new level of flexibility and control to industrial automation.

Understanding networking and communication in PLC systems is essential for any PLC programmer. It allows for the control of more complex systems and paves the way for more advanced automation scenarios.

Factors Influencing the Selection of Communication Protocols

Deciding on which communication protocol to use for your PLC system is a significant consideration. Here are some factors to consider when making this choice:

1. **Compatibility**: The protocol should be compatible with the PLCs you're using and any other devices you intend to connect to your system.
2. **Network Size and Complexity**: Different protocols are better suited to different sizes of networks. For example, Ethernet/IP can support large and complex networks, whereas Modbus might be more suitable for smaller, simpler systems.
3. **Data Requirements**: Different protocols support different data rates. Depending on the amount of data you need to transmit and the speed at which you need to transmit it, one protocol might be more suitable than others.
4. **Reliability**: Some protocols have built- in features for error detection and recovery, making them more reliable in critical applications.

The Role of Gateways in PLC Communication

In some cases, you might need to communicate between devices that use different protocols. This is where gateways come in. A gateway is a device that translates between different communication protocols, allowing devices that wouldn't be able to communicate with each other to do so.

For example, if you have a PLC using the Modbus protocol and a computer network using Ethernet/IP, you could use a Modbus to Ethernet/IP gateway to enable communication between the PLC and the computer network.

Wireless Communication

In some applications, it may not be possible or desirable to run network cables between your PLCs. In these cases, wireless communication can be used.

Many PLCs support wireless communication protocols such as Wi-Fi or Bluetooth. These can be useful in applications where the PLCs are moving (such as on a robot or vehicle) or where running network cables would be difficult or costly.

Keep in mind that while wireless communication brings a lot of convenience, it can also introduce new challenges, such as interference from other wireless devices, signal range limitations, and security concerns.

As you can see, networking and communication in PLC systems can be complex, but it's an essential part of modern PLC programming. By understanding these concepts, you'll be better equipped to design and implement effective PLC systems for a wide range of applications.

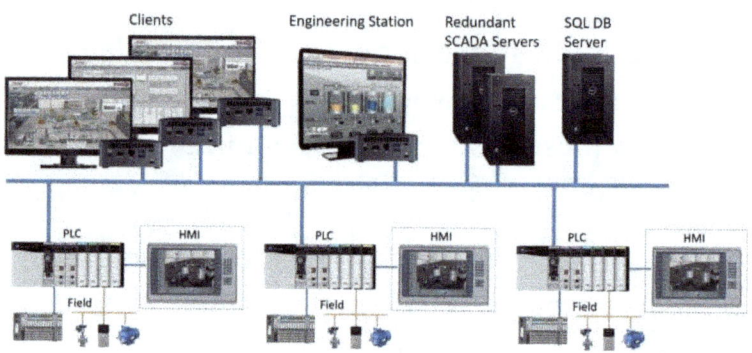

Redundancy in PLC Communication

Redundancy is a critical concept in PLC communication. In the context of PLC systems, redundancy refers to the duplication of critical components or functions in a system with the intention of increasing reliability.

Consider a large factory with a critical conveyor system. Suppose there's a single PLC controlling this system, and if it fails, the entire production halts. Here, a redundant PLC system would have a second, duplicate PLC ready to take over if the first one fails.
Communication between the two PLCs is crucial to synchronize their operations and ensure a smooth transition in the case of a failure.

Example of a PLC Communication

Consider a packaging plant where multiple PLCs control different parts of the process, such as sorting, filling, capping, and labeling. The sorter PLC sorts the items to be packaged and then signals the filler PLC to begin filling. Once filling is complete, the filler PLC sends a signal to the capper PLC to cap the package. This process continues down the line until the product is fully packaged and labeled.

Each of these steps involves communication between different PLCs, and the whole process can be controlled and monitored from a central SCADA system. Understanding how to program this communication is crucial to ensuring the smooth and efficient operation of the packaging plant.

Diagnostics in PLC Communication

Diagnostics is another critical aspect of PLC communication. PLCs often have built-in diagnostic capabilities that allow you to monitor the status of the PLC and its communication. For instance, you can check whether the PLC is sending and receiving messages correctly, the status of its network connections, and any errors that may have occurred.

Suppose you have a PLC system controlling a water treatment process. The PLC communicates with various sensors and actuators throughout the process, controlling pumps, valves, and other devices based on the sensor readings. If the PLC loses communication with a sensor, it can trigger an alarm to alert you to the problem. Additionally, you can log this information to help troubleshoot the issue.

With these enhanced concepts and examples, it should be clearer how pivotal the networking and communication aspect is in the realm of PLC programming. It helps achieve efficient, synchronized operations and safeguards against any sudden system failures.

Coming up in the next chapter, we'll discuss safety considerations in PLC programming, a segment that every programmer must be familiar with to maintain a secure operational environment.

Chapter 11: Safety Considerations in PLC Programming

As PLCs have evolved, so too have the safety standards and regulations guiding their usage. An improper command or an overlooked scenario in a PLC program can cause damage to machinery, halt production, or even lead to injuries. In this chapter, we will discuss key safety considerations that PLC programmers need to be aware of.

Designing for Fail-Safe Conditions

Designing a program for fail-safe conditions is paramount to ensure safety. Consider, for example, a PLC controlling a hydraulic press. If there's a power failure or a problem with the PLC itself, the press should return to its safe position to avoid any damage or injury. The safe state of a system is usually the state with the least amount of energy. For the hydraulic press, this would be the de- energized state where the press is open.

Routine Testing and Maintenance

Routinely testing and maintaining your PLC system can help prevent safety issues. This includes regular checks of PLC hardware, reviewing PLC programs for potential issues, and testing the fail-safe conditions of the system.

Handling of Emergency Situations

Most PLC-controlled systems should have provisions for handling emergencies. This could be a button or switch that puts the system into a safe state, such as an emergency stop (E-stop) button.

For example, consider a conveyor system with several emergency stop buttons placed along the conveyor's length. If someone presses an E-stop button, the PLC program should immediately stop the conveyor to prevent any potential injuries.

Understanding Safety PLCs

Safety PLCs are a specific type of PLC designed to meet higher safety standards.

They have redundant and fault-tolerant features, making them more reliable. These PLCs are often used in safety-critical applications, such as in the nuclear, aerospace, and automotive industries.

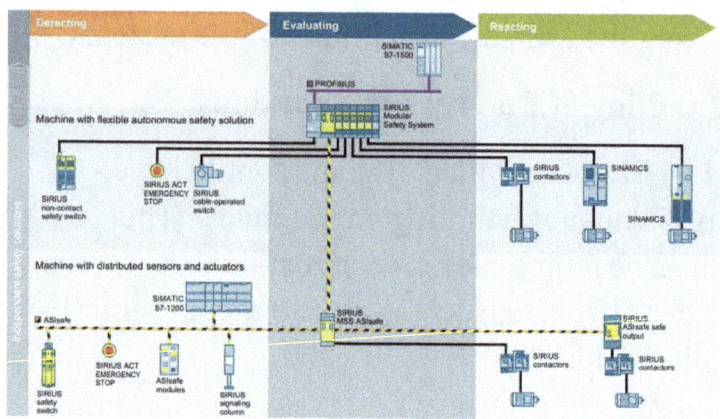

Consider, for instance, a safety PLC controlling a robot in a car manufacturing line. The PLC must ensure that the robot stops immediately if a human enters the robot's working area. The PLC must do this reliably, even in the event of failures or faults in the PLC or its input devices.

As PLC programmers, understanding these safety considerations will help you design and implement more reliable and safer systems. In the next chapter, we'll be looking at troubleshooting PLC systems, which is a critical skill for maintaining and ensuring the continued operation of these systems.

Safety Integrity Levels (SILs) in PLC Systems

Safety Integrity Level (SIL) is a measure of safety system performance, in terms of Probability of Failure on Demand (PFD). This is particularly crucial in the process industry. SIL levels range from SIL 1 (least reliable) to SIL 4 (most reliable). For example, a PLC system controlling an oil refinery's process might require a high SIL level because a failure could lead to serious consequences like an explosion.

Redundancy for Safety

For safety-critical applications, redundancy can be added to PLC systems to enhance their reliability. This might involve using redundant components, such as sensors, actuators, or even entire PLCs.

For instance, a fire suppression system might use redundant smoke detectors to confirm the presence of smoke before activating. This reduces the risk of false activations, which could damage equipment and disrupt operations.

Use of Safety-rated Devices

Many safety-critical systems require the use of safety-rated devices. These devices have been tested and certified to meet specific safety standards.

Consider a safety light curtain used to protect personnel working near hazardous machinery. If someone or something breaks the light curtain, a safety-rated PLC must reliably stop the machinery to prevent injury.

Examples of Fail-Safe Programming

Let's consider a real-world example to illustrate the importance of fail-safe programming. Assume you're programming a PLC for an elevator control system. The PLC receives input from buttons inside the elevator car, from buttons on each floor, and from sensors indicating the car's position.

A fail-safe condition here would be: if the PLC program crashes or loses power, the brakes should be applied to prevent the car from moving. If the door sensors indicate an object is blocking the doors, the doors should not close. These are fail-safe conditions because they default to the safest state in the event of a failure or abnormal condition.

Safety considerations in PLC programming are vital, as they prevent equipment damage, avoid production losses, and most importantly, protect people from injuries. As we move into the next chapter, we'll be discussing how to troubleshoot PLC systems—an essential skill in maintaining smooth and safe operations.

Safety Regulations and Standards

Different countries and industries have their own set of safety standards and regulations. For PLC programmers, it's crucial to understand and follow these standards when designing and implementing PLC systems.

For example, the International Electrotechnical Commission (IEC) provides a range of safety standards for PLCs, including IEC 61131-3 for PLC programming languages and IEC 61508 for functional safety of electrical/electronic systems.

Furthermore, industry-specific standards like ISO 13849-1 for machinery safety and IEC 62061 for safety of machinery - functional safety of safety-related electrical, electronic, and programmable electronic control systems, provide guidance on achieving safety in industrial automation.

Safety within the PLC Programming Cycle

Safety considerations are essential throughout the PLC programming cycle - from initial design, coding, testing, and maintenance.

In the design phase, hazards and risks must be identified, and safety requirements must be clearly defined. In the coding phase, programmers need to follow good coding practices, like providing clear documentation, structuring the program well, and implementing fail-safe conditions as discussed earlier.

In the testing phase, the program must be thoroughly tested under various scenarios to ensure that it responds correctly to all possible conditions, especially emergency situations. Lastly, during the maintenance phase, the PLC system should be regularly inspected and tested to ensure it continues to function safely and correctly.

Safety Certification for PLC Programmers

While not always a requirement, having safety certification can demonstrate a PLC programmer's competence in safety-related aspects of PLC systems. It can also be beneficial when seeking employment or promotions.

Organizations like the TÜV Rheinland Group and the International Society of Automation (ISA) offer certification programs focused on safety in automation and control systems. For instance, the Certified Automation Professional (CAP) and Certified Control Systems Technician (CCST) programs from the ISA cover a broad range of automation and control topics, including safety.

Safety in PLC programming is not just about programming practices but also about understanding regulations, maintaining systems, and continuously learning about safety developments in the field. By considering safety at every stage of the PLC programming cycle, you can help create a safer working environment.

Chapter 12: Troubleshooting PLC Systems

Troubleshooting is an essential skill for any PLC programmer, as problems can arise at any time, and efficient problem-solving can save time, effort, and resources.

Understanding PLC Error Codes

Most PLC systems provide error codes or status indicators to help identify issues. For instance, a flashing light on the CPU module might indicate a hardware fault, while a specific error code on the PLC's programming software could imply a problem with the program.

Take, for example, the Allen-Bradley SLC 500 PLC system. It provides troubleshooting information through an LCD panel on the CPU module. An "ERR" light on the panel means there's a major fault, and the fault type can be found by looking at the error code on the display.

Using PLC Diagnostics

PLC systems often have built-in diagnostic capabilities. These can provide real-time information about the PLC's status and help identify any issues. For example, a diagnostic function could show the status of all inputs and outputs, or it could provide information about memory usage.

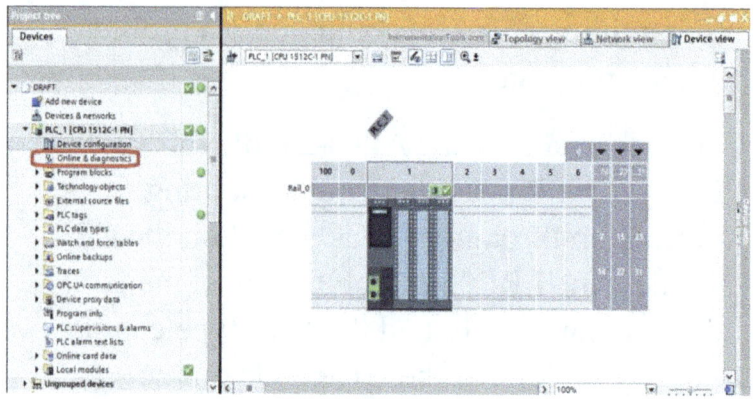

In Siemens S7 PLCs, for example, the diagnostic buffer keeps track of system events like startup, errors, and changes in mode. By understanding how to use such features, you can quickly identify and resolve issues.

Troubleshooting Common PLC Problems

There are some common problems that PLC programmers frequently encounter. These might include issues with power supply, input/output devices, communication systems, or the PLC program itself.

For instance, if an output device isn't responding as expected, you could check the status of the corresponding output in the PLC program. If the output is active in the program but the device isn't responding, the issue might be with the device itself, the wiring, or the output module.

Preventive Maintenance and Regular Testing

Preventive maintenance involves regularly checking the PLC system and performing necessary actions to prevent future problems. This might include cleaning dust from PLC modules, checking wiring connections, or updating software.

Regular testing is also crucial, as it can help identify potential problems before they become serious. For instance, you might perform a test of the PLC's response to a power failure to ensure that it recovers correctly.

Troubleshooting skills are an asset to any PLC programmer. By learning to interpret error codes, utilize diagnostic functions, troubleshoot common problems, and perform preventive maintenance and regular testing, you can ensure that your PLC systems continue to run smoothly and reliably.

Understanding the PLC Error Codes: Case Study

Taking our earlier example of the Allen-Bradley SLC 500 PLC system, imagine we encounter an error code on the LCD panel that reads "0001h." A look at the system's error code manual reveals that this represents a non-user fault during power up, meaning there is an issue with the power supply. In this case, we might start troubleshooting by checking the supply voltage level, verifying the power supply wiring, or even testing with a different power supply if one is available.

Using PLC Diagnostics: Real-world Example

Siemens S7 PLCs diagnostic buffer can be an extremely helpful tool. For example, suppose a factory's conveyor system suddenly stops. Checking the diagnostic buffer reveals an event that occurred at the same time as the stoppage: an emergency stop button was pressed. This allows us to immediately identify the cause of the problem, saving valuable time that would have been spent checking other parts of the system.

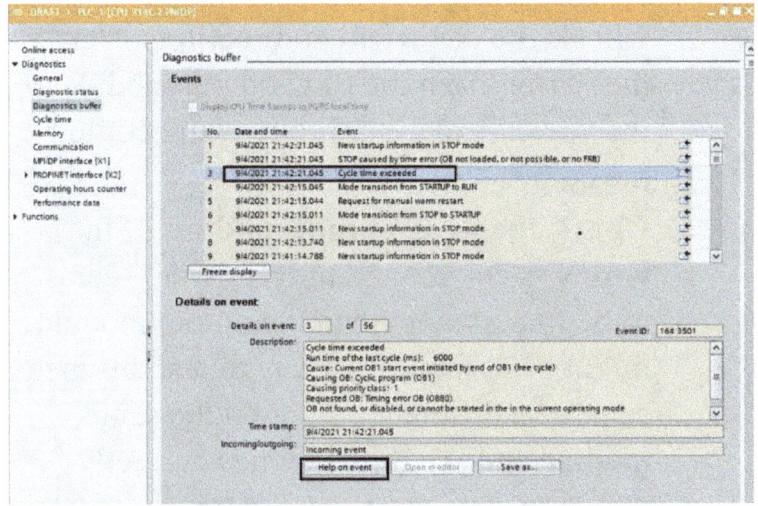

Troubleshooting Common PLC Problems: Scenario Analysis

Let's consider a real-world scenario. An automated assembly line stops due to a pneumatic cylinder not extending when expected. The PLC code appears to be activating the corresponding output, so what could be the issue?

Here's a systematic approach to this problem:

1. **Check the hardware**: Look at the physical cylinder and valve. Check if the valve is receiving the correct pneumatic pressure.
2. **Check the electrical connections**: Verify the wiring from the PLC output module to the valve. A loose or broken wire could be the culprit.
3. **Check the PLC output module**: If the wiring is correct and the hardware seems to be functioning, then the output module might be at fault. One way to test this is to activate a different output on the same module and see if it responds correctly.
4. **Check the PLC program**: If the above steps don't reveal the problem, there might be an issue with the PLC program. For instance, there could be a conflicting command elsewhere in the code that's preventing the output from activating.

Preventive Maintenance and Regular Testing: Further Insight

Preventive maintenance can avert many potential problems. For instance, dust can cause PLC modules to overheat, leading to premature failure. Regularly cleaning dust from modules can prevent this.

In terms of regular testing, it's important to simulate different scenarios that the PLC system could encounter and ensure it responds correctly. For instance, suppose a PLC system controls a safety door that must close when a specific alarm sounds. To test this, you could trigger the alarm in a controlled environment and check if the door closes as expected.

Troubleshooting PLC systems is a blend of system knowledge, hands-on skills, and systematic thinking. Understanding these principles in depth and applying them efficiently can make the process smoother and quicker.

PLC System Interaction

Observing the status of the PLC system directly can provide crucial information about the current state of the system. This can include monitoring the status LEDs on the PLC modules, using the human-machine interface (HMI) to view system status, or even using a laptop or other device to connect to the PLC and view real-time data. Rockwell Automation, the creators of Allen-Bradley PLCs, provides a thorough manual on interacting with their PLC systems, which you can find here: https://literature.rockwellautomation.com/idc/groups/literature/documents/um/1769-um021_-en-p.pdf

Creating Test Cases

Crafting test cases is an invaluable method to troubleshoot complex PLC programs. This includes defining a set of conditions that the program should meet and then checking if the program fulfills those conditions in all cases. Here's an example of a general troubleshooting checklist for PLC systems:

1. Check the PLC Input/Output (I/O) status: Is the hardware in the expected state?
2. Review the program logic: Is it designed to handle the current scenario?
3. Test for worst-case scenarios: Can the system handle maximum load?
4. Check for hardware faults: Is there an issue with the I/O modules, wiring, etc.?
5. Inspect the PLC error codes: Are there any error codes being displayed?

Leveraging Simulation Tools

Numerous PLC programming software packages incorporate simulation tools, which let you test your PLC program without needing to run it on actual hardware. These tools can identify logical errors in the program, such as endless loops, inactive outputs, or conditions that are never met. The Siemens TIA Portal software includes a PLC simulator named PLCSIM. Here's a link to

a tutorial on using the PLCSIM software:
https://new.siemens.com/global/en/products/automation/systems/industrial/plc/simatic-step-7-tia-portal.html

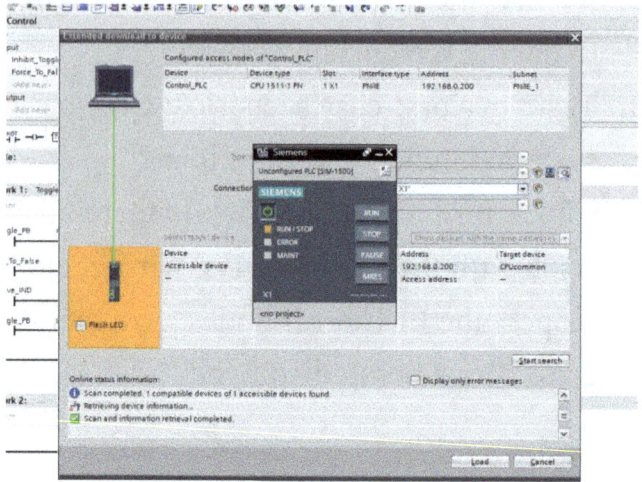

Utilizing Debugging Features

Most PLC programming software includes debugging features, which can help pinpoint and resolve issues in your program. These can include breakpoints, which pause the program at a specific point to inspect the current status of variables, inputs, and outputs, and step-through execution, which allows you to execute the program one instruction at a time to observe precisely how it behaves. Mitsubishi's GX Works2 software includes a debug mode that allows setting breakpoints, stepping through your program, and viewing real-time values of devices and variables. Here's a link to a guide on how to use these features: https://www.meau.com/eprise/main/sites/public/Downloads/Customer%20Center/Training%20PDFs/GX%20Works%202%20Beginner%27s%20Manual%20(Basic).pdf

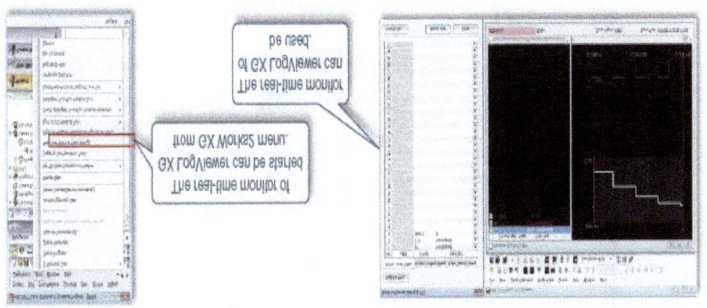

Mastering these advanced troubleshooting techniques and tools can help effectively diagnose and resolve issues in your PLC systems. Ensuring maximum uptime and performance of your automated systems and further establish your expertise as a proficient PLC programmer.

Step	Checklist Item	Details
1	Verify System Power	Check that the PLC system has power. This may seem basic, but it's often a common issue.
2	Check PLC Status	Look for any warning or error lights on the PLC, indicating a system issue. Refer to the PLC manual for the meaning of different light indicators.
3	Verify Input Status	Check the status of inputs in the PLC program. Are they what you would expect given the state of the physical inputs?

Step	Checklist Item	Details
4	Verify Output Status	Check the status of outputs in the PLC program. Are they what you would expect given the current program logic and input status?
5	Inspect PLC Code	Review the PLC code for any logical errors or potential issues. Use debugging features such as breakpoints and step-through execution to help with this.
6	Test Hardware	Test the physical PLC hardware if possible. This can include individual I/O

Step	Checklist Item	Details
		modules and communication devices.
7	Check Communication	Ensure that the PLC is communicating correctly with all networked devices. This can include other PLCs, HMIs, and network servers.
8	Review System Documentation	Consult any system documentation or manuals for potential solutions to the problem you're experiencing.
9	Perform Software Update	If all else fails, it may be beneficial to update the PLC's firmware to the latest version.

Step	Checklist Item	Details
		Always make sure to back up the current program before doing so.

Remember, safety should be the first priority when troubleshooting any electrical system, including PLCs. Always follow safe work practices and procedures.

Chapter 13: PLC Simulation and Testing

In any PLC project, an essential step is validating the program's functionality before implementation. This stage is called simulation and testing, where the software mimics the real-world operation, providing an opportunity to correct any logical errors before they cause serious problems.

Simulating PLC Logic

Simulating the PLC program involves running the PLC code in a controlled environment that can mimic the conditions it will encounter in the real world. This simulation environment can be as simple as a built-in feature of the PLC programming software, or as complex as a virtual 3D model of the system.

PLC programming software often includes built-in simulation tools. For instance, Rockwell's RSLogix 5000 has a feature known as 'Emulate' that allows for the simulation of a PLC program without requiring physical hardware. This is a powerful tool that can save time and resources during the initial stages of PLC programming.

It's also worth noting that some PLC systems support hardware-in-the-loop (HIL) simulation, where real-world devices can interact with the PLC software under test conditions. This method provides a highly realistic testing environment and can help reveal any unanticipated system interactions.

Functional Testing of PLC Program

Once the program has been implemented on the physical PLC, functional testing is the next crucial step. This involves running the program on the PLC itself and ensuring that it behaves as expected under normal conditions.

Functional testing can be carried out in a controlled environment, either by artificially triggering inputs to the PLC or by creating specific conditions within the real-world system. For example, if your PLC controls a conveyor belt system, you might test the program by manually activating the sensors that detect items on the belt, and verifying that the system reacts appropriately.

Stress Testing and Failure Modes

Stress testing is pushing the PLC system to its limits to ensure that it can handle the worst-case scenarios. This is crucial in industries where a minor malfunction can lead to significant consequences.

Simultaneously, PLC programs should be designed to handle failure modes. For instance, if a sensor fails, the PLC program should be able to identify this condition and react in a way that ensures the safety and integrity of the system.

Acceptance Testing

Finally, acceptance testing or user acceptance testing (UAT) is the final stage in the PLC programming process. This test is done in collaboration with the end-users of the system to ensure the system satisfies their requirements.

While thorough testing and simulation can never guarantee the absence of bugs or errors in a PLC program, it significantly reduces the risk of serious issues occurring during normal operation.

For detailed instructions on how to perform these tests using various PLC systems and software, Siemens provides a comprehensive guide on their website: https://support.industry.siemens.com/cs/document/109754502/testing-and-commissioning-in-tia-portal?dti=0&lc=en-WW

Remember, testing and simulating a PLC program are just as important as writing the program itself. A well-tested PLC program reduces the risk of serious problems occurring during normal operation, saving time, money, and potentially even lives.

In-Depth Analysis of Simulation Tools

Many PLC programming software have built-in simulation tools. For instance, 'PLCLogix' is a simulation software that replicates the functionality of the most popular PLCs in industry. It allows you to design, test, and debug PLC programs in an environment very close to the real one. Simulation software provides the ability to test your program with a wide range of 'what if' scenarios, which can prevent costly mistakes when the program is deployed in a live system.

Siemens' 'PLCSIM' and Rockwell's 'RSLogix Emulate' are other examples of such tools. They offer advanced simulation features such as creating a virtual controller, simulating I/O devices, and creating faults for testing PLC program reactions.

System-Level Testing: Beyond Individual Components

When testing the PLC program, it's important to look beyond individual inputs and outputs. System-level testing is about ensuring all components work together as intended. For example, a sensor might be functioning perfectly in isolation, but does it still perform adequately when the conveyor motor is running at full speed, causing electrical noise?

Creating such real-world scenarios during testing will make you confident about the reliability and robustness of the system.

Recovery Scenarios: Testing the Unforeseen

Stress testing is not just about overloading the system; it's about making sure the system can recover gracefully from unexpected scenarios. If a power outage occurs, does the system restart correctly when the power comes back? If a motor overheats and shuts down, does the PLC program identify this situation and take appropriate action?

These types of scenarios can be challenging to simulate, but they're critical to ensuring the robustness of the PLC program.

User Acceptance Testing: The Final Frontier

During acceptance testing, the PLC program is demonstrated to the end-users to validate the system functionality against the original requirements. But this is not just a yes/no validation. Often, the users can provide valuable feedback that can help to improve the system further.

For instance, the operators may suggest changes to the HMI layout to make it more intuitive. They may also recommend adding certain functionalities that could increase productivity or safety. Considering these recommendations can lead to a more successful project and increased user satisfaction.

Documenting the Process: A Key to Success

Last but not least, documenting the testing process and outcomes is critical. A well-documented test plan can provide a clear pathway for the testing process, while test reports can serve as a reference for future modifications or troubleshooting. Remember, a successful project is not just about making a system work, but also about creating a robust process that can be followed and improved upon in future projects.

Further details on performing PLC Simulation and Testing are available on various online resources, one of which is https://www.plcacademy.com/plc-programming-exercises/, which offers a plethora of information and exercises to understand the process better.

In the end, remember that a well-tested program not only ensures the system's reliability but also reduces downtime, increases safety, and saves costs in the long run.

Factory I/O: A 3D Simulator for PLC Training and Testing

In the realm of PLC simulation, Factory I/O stands out as a unique and highly practical tool. It's a 3D factory simulation for learning automation technologies. Designed to be easy to use, it allows to quickly pick up and play with little need for an in-depth knowledge of simulation.

Factory I/O uses a graphical interface where you can design and setup your virtual system by dragging and dropping components, and then connect them with your real or simulated PLC. The 3D design makes it easy to visualize the process and understand how the PLC will interact with the system. It supports a variety of industrial technologies including PLC, SCADA, PAC, and HMI.

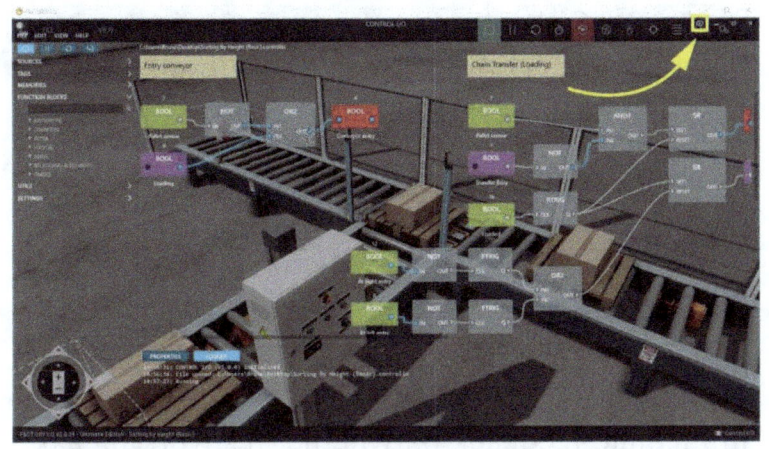

Making the Most of Factory I/O

With Factory I/O, you can create realistic industrial systems for training purposes or to prototype and test your PLC programs. It allows for implementing real-world scenarios with a vast array of industrial parts, including conveyor belts, elevators, and sorters, to mention a few.

Its interface with real PLCs and controllers is another strong advantage. This makes it a powerful tool for practical learning, letting you experiment in a safe environment before implementing in a real industrial setting.

It can communicate with most PLCs using standard protocols like OPC or Modbus, as well as PLC brands' own communication protocols like Siemens' S7 Protocol or Allen Bradley's Ethernet/IP.

Leveraging Factory I/O for PLC Testing

In the context of PLC testing, Factory I/O brings the advantages of immersive 3D visuals and real-time responses. You can see how your PLC program impacts the simulated factory and get immediate feedback on whether it's working as expected.

Moreover, it allows you to simulate various fault conditions and check how your PLC program responds to these situations. This can be a powerful way of troubleshooting your PLC program and making it more robust and reliable.

Factory I/O also allows you to observe the behavior of your system under different operating conditions, such as varying load or fluctuating power supply. This is an excellent way to ensure that your PLC program will perform consistently in the real world, regardless of external conditions.

Resources for Learning and Using Factory I/O

Real Games, the company behind Factory I/O, provides a range of tutorials and documentation to help users get started with their product. You can find these resources on the official Factory I/O website: https://factoryio.com/manual/. These resources include how- to guides, video tutorials, and case studies showing how Factory I/O is used in different industries.

Remember, the ultimate goal of simulation and testing is to build a PLC program that is efficient, reliable, and easy to maintain. By using tools like Factory I/O, you can get closer to this goal and ensure that your PLC program will stand up to the demands of the real world.

Chapter 14: Real-World PLC Programming Examples

PLC programming can vary greatly in complexity, depending on the specific application. In this chapter, we will explore two real-world PLC programming examples, one relatively simple and the other more complex. It is important to note that these are just examples, and actual industrial applications may require adjustments or additional features to suit specific needs.

Example 1: Conveyor Belt Control System

A simple yet common use case for PLCs in the manufacturing industry is to control a conveyor belt system. Let's consider a production line that moves items along several workstations. At each workstation, a task is performed on the item, such as assembly, inspection, or packaging.

In this case, the PLC program needs to control when the conveyor belt moves, stops, or changes speed, often depending on sensor inputs. For instance, a proximity sensor might detect when an item has reached a workstation, signaling the PLC to stop the conveyor belt. Once the task at the workstation is complete, another signal prompts the PLC to resume belt movement.

Example 2: Automated Bottling Plant

A more complex PLC application might involve an automated bottling plant. In this case, the PLC needs to control several processes, including bottle filling, capping, labeling, and packing into crates.

The PLC program coordinates these actions, often using sensor inputs and outputs to actuators. For example, a sensor might detect when a bottle is in position under the filling nozzle, and the PLC would then open the nozzle to fill the bottle. Once the sensor indicates that the bottle is full, the PLC would close the nozzle and signal the conveyor to move the bottle to the capping station.

This application could also involve conditional logic – for instance, if a sensor detects that a bottle is not capped correctly, the PLC could divert it to a rejection bin rather than moving it to the labeling station.

These examples illustrate the range of potential applications for PLCs and provide an insight into the sort of logic that a PLC program might need to implement. They should provide a useful foundation for your own PLC programming projects.

Remember, PLC programming is a complex task that requires a deep understanding of both the specific application and the PLC hardware and software. While these examples give a general sense of what PLC programming involves, they are only scratching the surface of this vast and fascinating field.

In the next chapter, we will discuss the future trends of PLC programming, as the field is continuously evolving with advancements in technology and changes in industrial requirements. For more real-world examples and in-depth PLC programming knowledge, please visit the official PLC manufacturer's website or community forums dedicated to PLC programming such as https://www.plctalk.net/qanda/.

As we continue to dive deeper into real- world PLC programming examples, it is important to understand that these examples are simplified scenarios. Actual industrial applications may be much more complex and require the integration of many more hardware components and sophisticated programming logic.

Example 3: Elevator Control System

Another example of PLC use is in elevator control systems. Here, the PLC must manage multiple inputs and outputs. Inputs can include floor selection buttons inside the elevator, calls from each floor, weight sensors to prevent overloading, and safety sensors on the doors. Outputs, on the other hand, could be signals to the motor controller, lights indicating the current floor or direction, and door open/close commands.

The PLC program must not only control the elevator's movement between floors but also handle exceptions. For example, if the elevator is full, the system should ignore calls from other floors until some passengers have exited. Or, if the doors are obstructed, the program needs to re-open the doors and prevent the elevator from moving.

Example 4: Traffic Light Control

Traffic light control is a classic example of PLC application in the field of transportation. PLCs are widely used to control the sequence of traffic lights at intersections, ensuring smooth and safe road traffic. The PLC program must cycle through a set sequence of light changes (green to amber to red and back to green) and manage the timings for each light. It may also need to take into account pedestrian crossing buttons or inputs from traffic sensors to adjust the timings during heavy or light traffic.

These real-world PLC programming examples demonstrate the broad application of PLCs and the diverse set of problems they can solve. The programming logic required in each case depends on the complexity of the system and the specific requirements of the application. As you can see, a well- programmed PLC is an incredibly powerful tool in managing and controlling automated systems.

To learn more about real-world PLC programming examples and explore further case studies, the https://www.rockwellautomation.com/ website provides a wealth of resources, and forums such as https://www.plctalk.net/qanda/ offer opportunities for discussion with experienced PLC programmers.

As we continue our journey into PLC programming, in the next chapter, we will focus on what the future holds for this exciting field. What are the upcoming trends and innovations that are set to shape the future of PLC programming? Let's find out in the following chapter!

Chapter 15: Future of PLC Programming: Trends and Predictions

In this final chapter, we will delve deeper into the future of PLC programming, exploring emerging trends and making predictions about the direction the field is headed. The rapid advancement of technology continues to shape the landscape of industrial automation, and PLC programming is at the forefront of these developments.

1. Internet of Things (IoT) Integration

The integration of PLCs with the Internet of Things (IoT) is a transformative trend that is expected to shape the future of PLC programming. IoT-enabled PLCs offer enhanced connectivity, real-time data exchange, and remote monitoring capabilities. With IoT integration, PLCs can seamlessly communicate with other devices, sensors, and systems, enabling more sophisticated automation and control processes. This integration will revolutionize industries such as manufacturing, energy, transportation, and building automation, making systems more efficient, flexible, and intelligent.

For example, in a smart manufacturing facility, PLCs can collect data from various sensors embedded in machines and send it to cloud-based analytics platforms. These platforms can then analyze the data to identify patterns, optimize production processes, and provide real-time insights for proactive decision-making.

2. Edge Computing and Distributed Intelligence

Edge computing is gaining prominence as a complementary technology to the cloud. In the context of PLC programming, edge computing refers to the deployment of computational capabilities closer to the physical devices and processes being controlled. This approach enables faster response times, reduced network latency, and enhanced data privacy. PLCs equipped with edge computing capabilities can process data locally, make real-time decisions, and communicate with other edge devices or the cloud as needed. This distributed intelligence brings greater autonomy and resilience to PLC systems, making them more adaptable and robust.

For example, in a smart grid application, PLCs installed at substations can analyze power consumption patterns, perform local load balancing, and quickly respond to grid disturbances, minimizing the need for centralized control and reducing dependency on a stable network connection.

3. Artificial Intelligence (AI) and Machine Learning (ML)

Artificial Intelligence (AI) and Machine Learning (ML) are poised to have a significant impact on PLC programming. These technologies empower PLCs to learn from data, identify patterns, and make intelligent decisions. By applying ML algorithms, PLCs can optimize control strategies, detect anomalies, and predict system failures. For example, predictive maintenance can be implemented, where the PLC uses ML models to analyze sensor data and identify patterns indicating potential equipment failures. This enables proactive maintenance, minimizing downtime and reducing maintenance costs.

In manufacturing, AI and ML can be leveraged to optimize production processes by dynamically adjusting control parameters, predicting quality issues, and optimizing energy consumption. PLCs can learn from historical data, adapt to changing operating conditions, and continuously improve system performance.

4. Cybersecurity and Resilience

As PLC systems become more interconnected and accessible through network connections, ensuring robust cybersecurity measures is crucial. The future of PLC programming will prioritize cybersecurity considerations to protect against cyber threats, unauthorized access, and data breaches. PLC programmers will implement advanced encryption algorithms, authentication mechanisms, and security protocols to safeguard the integrity and confidentiality of PLC systems. Additionally, efforts will focus on enhancing the resilience of PLC systems, enabling quick recovery from cyberattacks or system failures.

Secure communication protocols, network segmentation, and intrusion detection systems will be implemented to ensure the integrity of PLC programs and protect critical infrastructure from potential cyber threats.
Continuous monitoring and regular security audits will be performed to identify vulnerabilities and apply timely patches and updates.

5. Human-Machine Collaboration

The future of PLC programming will emphasize human-machine collaboration, where humans and machines work together harmoniously. This collaboration will leverage the strengths of each, with humans providing strategic decision-making, creative problem-solving, and overall system oversight, while machines handle repetitive tasks, data processing, and real-time control. Advanced human-machine interfaces, augmented reality (AR), and natural language processing will facilitate intuitive interaction with PLC systems, enabling operators and technicians to efficiently monitor, control, and troubleshoot complex processes.

For example, with the help of augmented reality, technicians can wear smart glasses that overlay real-time PLC data, diagnostics, and instructions onto their field of vision.
This enables them to quickly identify issues, access relevant documentation, and perform maintenance tasks with greater accuracy and efficiency.

6. Continued Advancements in Programming Environments

PLC programming environments will continue to evolve, offering more intuitive, user-friendly interfaces and tools. Graphical programming languages, such as function block diagrams and ladder logic, will be complemented by high-level languages, enabling programmers to choose the most suitable approach for each application.
Integrated development environments (IDEs) will provide comprehensive features for

debugging, code analysis, simulation, and collaboration.

In addition, the emergence of code generation and model-based development approaches will simplify PLC programming by allowing programmers to design control systems using visual models and automatically generating PLC code. This accelerates development cycles, improves code quality, and promotes code reusability.

Conclusion

The future of PLC programming holds immense potential, driven by advancements in IoT integration, edge computing, AI/ML, cybersecurity, human-machine collaboration, and programming environments. These trends will shape the way industries harness the power of PLCs for automation, control, and optimization. As PLC programmers, it is crucial to stay updated with these emerging technologies and adapt to new programming paradigms to unlock the full potential of PLC systems in the future.

Thank you for joining us on this journey through "The PLC Programming Guide for Beginners." We hope this book has provided you with a solid foundation in PLC programming concepts, techniques, and real-world applications. As you continue to explore and expand your knowledge in this field, remember to embrace the ever-evolving nature of PLC programming and embrace lifelong learning.

Wishing you success in your PLC programming endeavors!

Chapter 16: Navigating the Path to a PLC Programming Career

Congratulations on completing "The PLC Programming Guide for Beginners"! By now, you have gained a solid understanding of PLC programming and its applications. In this final chapter, we will discuss how to land a job working with PLCs, providing you with guidance and tips to kickstart your career in this exciting field.

1. Develop a Strong Foundation

Building a strong foundation in PLC programming is crucial for landing a job in the field. Ensure you have a solid understanding of PLC concepts, programming languages, and industrial automation processes. Review the topics covered in this book, and consider further education or training to deepen your knowledge. Online courses, certifications, and practical projects can help demonstrate your skills to potential employers.

2. Gain Practical Experience

Employers highly value practical experience in PLC programming. Seek opportunities to work on real-world projects, whether through internships, cooperative education programs, or volunteering. This hands-on experience will allow you to apply your knowledge in a practical setting and demonstrate your ability to troubleshoot, program, and optimize PLC systems.

Consider approaching local manufacturers, engineering firms, or system integrators for internship or entry-level positions. Even if these opportunities are unpaid or temporary, they provide valuable experience and a chance to network within the industry.

3. Build a Portfolio

Create a portfolio showcasing your PLC projects and programming skills. Include documentation, program samples, and project descriptions. If possible, highlight projects that align with the industry or application you are interested in, such as manufacturing, process control, or building automation. A well-organized portfolio demonstrates your ability to apply PLC programming principles and gives potential employers insight into your capabilities.

4. Network in the Industry

Networking plays a crucial role in finding job opportunities. Attend industry events, seminars, trade shows, and conferences related to automation and control systems.

Engage with professionals in the field, join online communities and forums, and connect with potential mentors. Networking can lead to valuable connections, job leads, and insights into the industry.

5. Customize Your Resume and Cover Letter

When applying for PLC programming positions, tailor your resume and cover letter to showcase your relevant skills and experiences. Highlight your PLC programming knowledge, specific projects you have worked on, and any certifications or training you have completed. Emphasize your problem-solving abilities, attention to detail, and familiarity with industry standards.

6. Prepare for Interviews

Prior to interviews, research the company and the industry it operates in. Familiarize yourself with common PLC applications, industry trends, and emerging technologies. Be prepared to discuss your technical knowledge, problem-solving abilities, and project experiences. Demonstrate your enthusiasm for the field and your willingness to learn and adapt to new challenges.

7. Continuous Learning and Professional Development

PLC programming is a rapidly evolving field, and employers value individuals who stay up to date with the latest technologies and advancements. Commit to continuous learning by following industry publications, participating in webinars, and exploring further training opportunities. Stay informed about new programming languages, hardware platforms, and software tools relevant to PLC programming.

1. Develop a Strong Foundation

To land a job working with PLCs, it's essential to have a strong foundation in PLC programming. Here are some key steps to develop your skills:

- Review PLC Concepts: Ensure you have a solid understanding of PLC fundamentals, such as digital and

analog I/O, ladder logic, programming languages, and communication protocols.
- Explore Industrial Automation: Familiarize yourself with different industrial automation processes and systems commonly used in the industry.
- Take Online Courses: Consider enrolling in online courses that focus on PLC programming. Platforms like Udemy, Coursera, and LinkedIn Learning offer comprehensive PLC programming courses.
- Earn Certifications: Pursue certifications relevant to PLC programming, such as the Certified PLC Technician (CPLT) or Certified Automation Professional (CAP) certifications.

2. Gain Practical Experience

Practical experience is highly valued by employers in the field of PLC programming. Here's how you can gain hands-on experience:

- Seek Internship Opportunities: Approach local manufacturers, engineering firms, or system integrators for internship positions. Even short-term or unpaid internships can provide valuable exposure to real- world PLC projects.
- Cooperative Education Programs: Explore cooperative education programs offered by universities or technical colleges. These programs allow you to alternate between classroom learning and work experience related to your field of study.
- Volunteer or Contribute to Open- Source Projects: Look for volunteering opportunities or contribute to open-source projects that involve PLC programming. This allows you to collaborate with others and gain practical experience.

3. Build a Portfolio

Creating a portfolio is an effective way to showcase your skills and projects to potential employers. Consider the following tips when building your portfolio:

- Document Your Projects: Document your PLC projects with detailed descriptions, including the objectives, programming languages used, system architecture, and notable challenges faced.
- Include Sample Code: Provide snippets of your PLC program code to demonstrate your programming proficiency.
- Visuals and Media: Incorporate visuals such as diagrams, screenshots, or videos that illustrate the functioning of your PLC projects.
- Highlight Project Outcomes: Discuss the outcomes and impact of your projects, such as increased productivity, cost savings, or improved efficiency.

- Tailor Your Portfolio: Customize your portfolio to align with the industry or application you are interested in. Highlight projects that demonstrate your skills and knowledge in those specific areas.

4. Network in the Industry

Networking plays a crucial role in finding job opportunities within the PLC programming field. Consider these strategies to expand your professional network:

- Attend Industry Events: Participate in industry events, trade shows, conferences, and seminars focused on automation, control systems, or PLC programming. These events provide opportunities to connect with professionals and potential employers.
- Join Online Communities and Forums: Engage with online communities and forums dedicated to PLC programming. Platforms like LinkedIn, Reddit, or specialized PLC programming forums

allow you to connect with experienced professionals, ask questions, and learn from their insights.
- Seek Mentors: Identify professionals in the field who can mentor you and provide guidance on your career path. Mentors can offer valuable advice, share their experiences, and provide recommendations or referrals.

5. Customize Your Resume and Cover Letter

When applying for PLC programming positions, it's important to customize your resume and cover letter to make a strong impression. Consider the following tips:

- Highlight Relevant Skills: Emphasize your PLC programming skills, including knowledge of programming languages (such as ladder logic, structured text, or function block diagrams), HMI/SCADA systems, communication protocols (such as Modbus or Ethernet/IP), and

experience with PLC hardware and software platforms.
- Showcase Project Experience: Highlight projects you have worked on, mentioning the objectives, technologies used, and outcomes achieved. Quantify the results wherever possible, such as improved production efficiency or reduced downtime.
- Certifications and Training: Include any relevant certifications, training courses, or workshops you have completed. This demonstrates your commitment to professional development and staying up to date with industry trends.

6. Prepare for Interviews

Preparing for interviews is crucial to showcase your skills and knowledge effectively. Here are some areas to focus on:

- Technical Knowledge: Review PLC programming concepts, programming languages, and industrial automation processes. Be prepared to discuss how

you have applied these concepts in your projects.
- Problem-Solving Abilities: Highlight your problem-solving skills, particularly in troubleshooting PLC systems or optimizing control strategies. Prepare examples of challenging situations you have encountered and how you successfully resolved them.
- Project Experience: Be ready to discuss your project experience in detail, explaining the objectives, challenges faced, and the impact of your contributions.
- Familiarity with Industry Trends: Stay updated on emerging technologies, industry standards, and trends in PLC programming. Be prepared to discuss how you stay informed and how you adapt to new developments.

7. Continuous Learning and Professional Development

To excel in the field of PLC programming, continuous learning and professional development are key. Consider these resources and strategies:

- Industry Publications: Subscribe to industry publications such as "Automation World," "Control Engineering," or "IEEE Transactions on Industrial Electronics" to stay informed about the latest developments and trends in PLC programming.
- Webinars and Online Courses: Attend webinars and enroll in online courses related to PLC programming. Websites like Udemy, Coursera, and LinkedIn Learning offer a wide range of courses on automation, control systems, and PLC programming.
- Professional Associations: Join professional associations such as the International Society of Automation (ISA) or the Institute of Electrical and Electronics Engineers (IEEE). These organizations offer networking opportunities, webinars, conferences,

and resources to support your professional development.
- Blogs and Online Forums: Follow influential blogs and participate in online forums dedicated to PLC programming. Platforms like PLCdev, PLC Talk, or MrPLC provide valuable insights, discussions, and tips from experienced professionals.

Conclusion

By developing a strong foundation, gaining practical experience, building a portfolio, networking, customizing your resume, preparing for interviews, and embracing continuous learning, you can increase your chances of landing a job working with PLCs. The field of PLC programming offers diverse career paths and exciting opportunities for growth. Remember to stay enthusiastic, open to new challenges, and adaptable to the evolving technology landscape. Best of luck as you embark on your journey to a rewarding career in the world of PLCs!

Resume Example 1:

John Smith 123 Main Street Anytown, USA Phone: (555) 123-4567 Email: johnsmith@email.com

Objective: Highly skilled PLC programmer with extensive experience in industrial automation. Seeking a challenging position where I can utilize my programming expertise to optimize control systems and drive operational efficiency.

Education: Bachelor of Science in Electrical Engineering ABC University, Anytown, USA Graduated: May 20XX

Certifications:

- Certified PLC Technician (CPLT)
- Siemens Certified Programmer (SCP)

Skills:

- Proficient in ladder logic, structured text, and function block diagrams
- Experience with PLC hardware and software platforms, including Siemens and Allen-Bradley

- Strong knowledge of communication protocols such as Modbus, Ethernet/IP, and Profibus
- Familiarity with HMI/SCADA systems and data acquisition
- Skilled in troubleshooting, debugging, and optimizing PLC systems
- Excellent problem-solving abilities and attention to detail
- Strong verbal and written communication skills

Experience: PLC Programmer XYZ Manufacturing Company, Anytown, USA May 20XX - Present

- Developed and implemented PLC programs to optimize production processes, resulting in a 20% increase in efficiency.
- Conducted system analysis and troubleshooting to resolve issues and minimize downtime.
- Collaborated with cross-functional teams to integrate PLC systems with

HMI/SCADA interfaces for real-time monitoring and control.
- Led the commissioning of PLC systems for new production lines, ensuring seamless integration and successful operation.

Project Highlights:

- Implemented predictive maintenance strategies using machine learning algorithms, reducing equipment failure rates by 15%.
- Designed and programmed a control system for a robotic assembly line, resulting in a 30% improvement in cycle time.

Resume Example 2:

Emily Johnson 456 Elm Street Anytown, USA Phone: (555) 987-6543 Email: emilyjohnson@email.com

Objective: Results-driven PLC programmer with a passion for optimizing industrial processes. Seeking a challenging position where I can leverage my programming skills and project management experience to drive operational excellence.

Education: Associate Degree in Automation and Control Systems DEF Technical Institute, Anytown, USA Graduated: May 20XX

Skills:
- Proficient in ladder logic and structured text programming languages
- Experience with PLC hardware and software, including Rockwell Automation and Schneider Electric platforms
- Knowledge of industrial communication protocols such as Ethernet/IP, Profinet, and DeviceNet
- Strong understanding of PID control loops and process optimization techniques
- Skilled in HMI/SCADA design and configuration
- Excellent problem-solving and analytical skills

- Effective written and verbal communication abilities

Experience: Junior PLC Programmer ABC Engineering Services, Anytown, USA June 20XX - Present

- Assisted senior programmers in developing PLC programs for manufacturing and process control systems.
- Conducted system testing and troubleshooting to ensure optimal performance and functionality.
- Collaborated with engineering teams to design HMI interfaces and implement data acquisition systems.
- Assisted in the installation, commissioning, and startup of PLC systems at client sites.

Project Highlights:

- Contributed to the development of a PLC-based control system for a water treatment plant, resulting in a 25% reduction in energy consumption.

- Implemented a fault detection and diagnostic algorithm for a HVAC system, leading to a 15% improvement in system reliability.

Cover Letter Example 1:

Dear Hiring Manager,

I am writing to express my interest in the PLC Programmer position at XYZ Company. With a Bachelor's degree in Electrical Engineering and over five years of experience in industrial automation, I am confident in my ability to contribute to your team.

In my current role as a PLC Programmer at ABC Manufacturing Company, I have successfully developed and implemented PLC programs to optimize production processes, resulting in significant efficiency gains. I am proficient in ladder logic, structured text, and function block diagrams, and have experience working with PLC hardware and software platforms, including Siemens and Allen-Bradley. Additionally, my strong knowledge of communication protocols such as Modbus, Ethernet/IP, and Profibus enables me to effectively integrate PLC systems with other devices and systems.

I am a detail-oriented problem solver with a passion for optimizing control systems. I thrive in collaborative environments and have a proven track record of delivering projects on time and within budget. I am confident that my skills and experience make me an ideal candidate for this position.

Thank you for considering my application. I look forward to the opportunity to discuss how my skills and expertise align with the needs of XYZ Company. Please find my attached resume for your review.

Sincerely, John Smith

Cover Letter Example 2:

Dear Hiring Manager,

I am excited to apply for the PLC Programmer position at ABC Engineering Services. With an Associate Degree in Automation and Control Systems and hands-on experience in PLC programming, I am confident in my ability to contribute to your projects and drive operational excellence.

In my current role as a Junior PLC Programmer, I have been involved in the development and testing of PLC programs for manufacturing and process control systems. I am skilled in ladder logic and structured text programming languages and have experience with Rockwell Automation and Schneider Electric platforms. My understanding of industrial communication protocols such as Ethernet/IP, Profinet, and DeviceNet enables me to effectively integrate PLC systems in diverse environments.

I am passionate about optimizing industrial processes and ensuring the smooth operation of control systems. I am a quick learner with excellent problem-solving skills, and I am always eager to expand my knowledge in the field. I believe that my technical expertise, combined with my strong communication and collaboration abilities, make me a valuable asset to your team.

Thank you for considering my application. I have attached my resume for your review. I would welcome the opportunity to discuss how my skills and experience align with the needs of ABC Engineering Services.

Sincerely, Emily Johnson

Here is a list of relevant interview questions to help you prepare for a PLC programming position:

1. Can you explain the basic components of a PLC system and their functions?
2. What programming languages are commonly used in PLC programming, and what are their advantages and disadvantages?
3. How do you troubleshoot and debug PLC programs when encountering issues?
4. Can you describe a complex project you worked on involving PLC programming? What challenges did you face, and how did you overcome them?
5. How do you approach optimizing a PLC control system to improve efficiency and productivity?
6. What are the key considerations when integrating a PLC system with HMI/SCADA interfaces or other external devices?

7. Can you explain the importance of communication protocols in PLC programming and give examples of protocols you have worked with?
8. How do you ensure the security and integrity of a PLC system, especially in relation to potential cyber threats?
9. Have you worked on a project that required compliance with specific industry standards or regulations? How did you address those requirements?
10. Can you discuss your experience with PLC networking and communication? What protocols and techniques have you used?
11. How do you approach documenting your PLC programs and maintaining clear and organized project files?
12. Have you worked in a team environment on PLC projects? How did you collaborate with other team members to achieve project goals?
13. What methods do you use to keep up to date with the latest trends

and advancements in PLC programming?
14. Can you provide an example of a situation where you had to troubleshoot a complex PLC system and how you resolved the issue?
15. How do you ensure the safety of operators and personnel when designing and programming PLC systems?

Remember to prepare thoughtful and concise answers for these questions based on your experiences and knowledge.

Additionally, be ready to discuss specific projects, challenges, and outcomes that highlight your expertise and problem-solving abilities in PLC programming. Good luck with your interview preparation!

www.ingramcontent.com/pod-product-compliance
Lightning Source LLC
Chambersburg PA
CBHW072138170526
45158CB00004BA/1427